EXPLORING
THE
UNIVERSE

Robin Kerrod

Silverdale Books

Published by SILVERDALE BOOKS
An imprint of Bookmart Ltd
Registered number 2372865
Trading as Bookmart Ltd
Blaby Road
Wigston
Leicester LE18 4SE

© 2006 Graham Beehag Books

Graham Beehag Books
Christchurch
Dorset BH23 8BN

e-mail books@freeuk.com

This edition published 2007

ISBN 978-1-845094-34-8

Design and editorial: Graham Beehag

Printed in Singapore

3 5 7 9 10 8 6 4 2

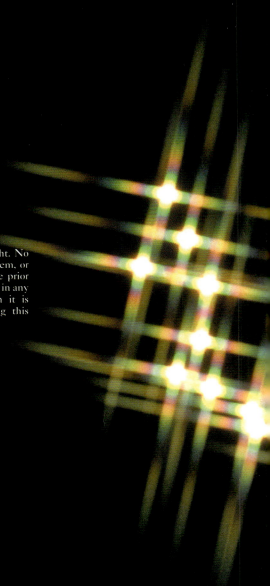

Contents

The Galoleo probe is pictured here flying by the amazing volcanic moon Io, as it orbits the giant planet Jupiter.

Exploring the Solar System

The Sun hurtles through space with a huge family of bodies large and small — planets, moons, asteroids, comets, and meteoroids. We call this family of the Sun the solar system.

The planets form the most important part of the solar system. There are nine planets in all, which circle at different distances from the Sun. They are Mercury, Venus, Earth, Mars, Jupiter, Saturn, Uranus, Neptune, and Pluto.

Many of the planets have smaller bodies circling round them. We call them satellites, or moons. The Earth has one satellite, the Moon. Some planets have many moons — Saturn and Uranus each have at least 18.

There are many other bodies in the solar system besides the planets and their moons. There are millions of rocky lumps we call asteroids, billions of iceballs we call comets, and countless swarms of little bits we call meteoroids.

Astronomers have been studying the solar system for thousands of years, first with their eyes, then through telescopes, and now through the telescopes and cameras of spacecraft. Space satellites keep an eye on the heavenly bodies from Earth orbit. Space probes journey for billions of miles through space to observe them from close quarters. The exciting discoveries these spacecraft have made have revolutionized our knowledge of the solar system.

THE SUN'S FAMILY

The Sun lies at the heart of the solar system. Its gravity hols in place the huges family of bodies, large and small, that make up the solar system. The Sun and its family were born billions of years ago out of a huge cloud of gas.

Every day at dawn we see the Sun rise above the horizon in the east. Moving westward, it arcs through the sky during the day before setting in the west in the evening. It seems that the Sun circles around Earth.

And that is what ancient astronomers believed. They thought that Earth was the centre of the universe, and that the Sun and all the heavenly bodies circled around Earth.

Few people questioned this view until 1543. In that year, a Polish churchman and astronomer named Nicolaus Copernicus suggested that it was Earth that circled around the Sun, and not the other way round. The other planets circled around the Sun, too. Earth was just another planet.

Copernicus was right. Earth and the planets all circle around the Sun in space. They belong to a Sun-centred, or solar system, not an Earth-centred system.

The Sun dominates our part of space. It is quite a different body from the planets and all the other bodies in the solar system. It is a star. It is just like the stars we see in the night sky but very much closer. It is a great ball of very hot gas that gives out vast amounts of energy as light and heat.

In contrast, the other bodies in the solar system are made up of rock, ice, or cold gas. And they give out no light of their own. We see them shining in the night sky only because they are lit up by the Sun, and reflect the sunlight into our eyes.

The Sun holds the planets and other bodies in place with its enormous gravity, which stretches out billions of miles into space. It has an enormous gravity because of its huge mass — it has 750 times the mass of all the other bodies in the solar system put together.

Opposite: Moments before sunset over the ocean.

Left: Craters scar Saturn's moon Tethys, one of more than 60 moons in the solar system.

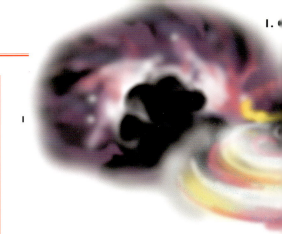

Birth of the Solar System

The vast space between the stars is almost completely empty, but not quite. It contains minute traces of gas and dust. In places, the gas and dust gather together to form vast billowing clouds, which astronomers call nebulae. It is in such clouds that stars are born.

Five billion years ago, there was a huge nebula in our part of space. There was nothing else — no Sun, no planets, no asteroids, no comets — just an enormous dark cloud of cold gas and dust. Then one day gravity came into play. In part of the cloud, whiffs of gas and swarms of tiny particles of dust began attracting one another. The cloud began to collapse, or shrink, and become denser.

Gradually, over many thousands of years, the cloud turned into a rough ball-shape and started to spin round. At the same time it began warming up. As the ball continued to collapse, its centre part became densest and hottest and began to glow feebly. The more the ball of matter collapsed, the faster it

spun round. This caused the cooler material in the outer regions to form into a fat disk.

Over time, the glowing ball at the centre became smaller and hotter, and the surrounding disk of gas and dust got thinner and thinner.

When temperatures in the core, or heart, of the hot ball reached millions of degrees, a dramatic change took place. The centres, or nuclei, of atoms of hydrogen gas began to fuse, or join together. This nuclear reaction produced vast amounts of energy, which poured out into space as heat and light. The hot ball had turned into a star — a star we now call the Sun.

The planets are born

The young Sun poured out light and streams of particles into the disk of matter that surrounded it. They began blowing away the gases from the warmer inner part of the disk into the colder outer regions. Many of these gases were frozen into icy lumps.

In the colder outer parts of the disk, bits of rock and ice began bumping into one another and sticking together to form bigger lumps. Later they began to gather the gases around them to form larger bodies. In this way, the gas giant planets, from Jupiter through Neptune, came into being.

The Orion nabula, stars are born all the time. It is one of the closest star-forming regions in the heavens.

8

2. and 3. Gas cloud begins to shrink and spin around

3

In the warm inner part of the disk, bits of rocks and metals also began smashing into one another and sticking together to form larger and larger lumps. Over time the lumps became the great masses we know as the inner planets, from Mercury through Mars.

Many lumps of matter were left over after the planets formed. Today we find some — the asteroids — circling in a great ring between the orbits of Mars and Jupiter. Others — mainly comets — we find at the very edge of the solar system.

4. The centre heats up and over time becomes the Sun

4

5

5. and 6. Matter in the surrounding disk clumps together to form the planets

6

Other Solar Systems

There are many other stars in the universe like the Sun — billions in our own galaxy alone. Almost certainly, these other distant suns form in the same way as our Sun. This presumably means that disks of matter form around them too, and from these disks planets eventually appear. In which case, there should be plenty of other solar systems in space, similar to our own.

Until recently, no one was sure if other solar systems did exist. But recent findings suggest that they do. The Hubble Space Telescope has sent back images of disks of matter around newly formed stars. And astronomers have found convincing evidence of extrasolar planets — planets around other stars.

Humans have long speculated that strange forms of life may lurk on other planets among the stars.

Tell-tale wobbles

One way astronomers search for extrasolar planets is by checking the way stars move. If they see a star wobble slightly in a regular way, they reckon that it is being pulled out of line by the gravity of one or more circling planets.

Astronomers spent years developing instruments and techniques to detect the minute wobbles in star motions before making their first discoveries in 1995. They found a planet about half the size of Jupiter orbiting around a Sun-like star called 51 Pegasi.

Since then many extrasolar planets have been discovered. Most seem to be similar in size to Saturn or Jupiter, although some are bigger. It would need large planets like these to make a star wobble enough to be detected. Late in 1999, astronomers made the first sighting of an extrasolar planet through a telescope, around the star Tau Bootis. It seems to be bluish in colour and nearly twice as big as Jupiter.

Life on another Earth?

If there are millions of other solar systems in space, the chances are that there are some like our own. So, presumably, somewhere there must be planets on which conditions are the same as on Earth.

The question now arises: is there life on these other Earth-like planets? This is difficult to answer because we don't really know how life began on Earth. Most scientists think that life on Earth had a chemical origin. Billions of years ago certain carbon chemicals began reacting together to form more complicated substances. Over time, these substances began to find ways of reproducing themselves, and life began.

If life did begin chemically like this, there should be plenty of life elsewhere in the universe. We call this life extraterrestrial, meaning not of Earth. And there should also be intelligent extraterrestrial beings, which we popularly call ETs or aliens.

The question arises: where are they and why haven't we heard from them? Because the stars lie light-years away, any signals from alien civilizations would take years — maybe centuries — to reach us, even if they were sent in our direction. Nevertheless, astronomers on Earth are using radio telescopes to listen for likely signals from other civilizations among the stars.

The powerful radio telescope in Puerto Rico has sent coded messages **(above)** into deep space for any extererrestrial who might be listening.

COMPARING THE PLANETS

The nine planets of our solar system have several things in common: they all follow much the same path through the heavens; they all circle the Sun in the same direction; and they all spin on their axis in space. But in size and make-up, the planets are very different indeed.

Just like ancient astronomers, we can see five "wandering stars", or planets, with the naked eye. But we can hardly ever see them all at the same time. They appear at different times and in different parts of the sky. And they vary widely in brightness.

The easiest planet to spot is Venus. On many nights of the year, it appears in the western sky just after sunset. Then we call it the evening star. It appears a long time before the real stars come out. At other times of the year, if we get up early, we can see Venus shining in the eastern sky before sunrise. Then we call it the morning star. Morning or evening, Venus easily outshines all the stars and all the other planets.

Mercury too can be a morning or an evening star. But it is always difficult to see, because it stays close to the Sun and is always low down on the horizon.

After Venus, Jupiter is the planet that shines brightest. It is usually seen in the dark night sky and is impossible to miss, shining like a beacon among the fainter stars.

Mars can sometimes become as bright as Jupiter, but usually it is fainter. When they are both bright, it is easy to tell which is which. Jupiter shines with a brilliant white light, but Mars looks reddish because its surface is red. This is why we call it the Red Planet.

Saturn is the farthest of the planets we can see with the naked eye, and never gets as bright as the others. Usually it is not much brighter than the stars. This makes it difficult to spot unless you know exactly where to look. Saturn would look even dimmer were it not for its magnificent rings.

Below: The surface of Mars. Mars is one of the four rocky planets located in the centre of the solar system.

Jupiter is sometimes called the king of planets. This great ball of gas and liquid is bigger than all the other planets put together. The small body seen in front of it is its colourful moon Io.

How the Planets Move

From everywhere on Earth, it appears as if the night sky is a great dome above our heads. It is as if Earth is in the middle of a great dark ball, or sphere. Ancient astronomers believed this, and they called it the celestial sphere. They thought that the stars were stuck on the inside of the sphere.

Earth circles around the Sun once a year. But from Earth it seems that the Sun travels around the celestial sphere — through the stars — during the year. It seems to travel through a number of star patterns, or constellations. They are called the constellations of the zodiac.

Earth and the planets circle the Sun in much the same plane. This means that from Earth, the planets are always found close to the path of the Sun in the sky. In other words, they are always found in the constellations of the zodiac.

Day by day

Like all the heavenly bodies, Earth moves through space in two ways. It not only travels in its orbit around the Sun.

Venus shines brightly in the evening sky on many nights of the year. Here it is seen close to the Moon.

It also spins round in space like a top. We say it spins on its axis, which is an imaginary line through its North and South Poles. The planets also spin on their axes in a similar way as they travel in orbit around the Sun.

Earth spins round once in space in 24 hours. This is the period of time we call a day. All the planets have different spin periods, or "days". The giant planet

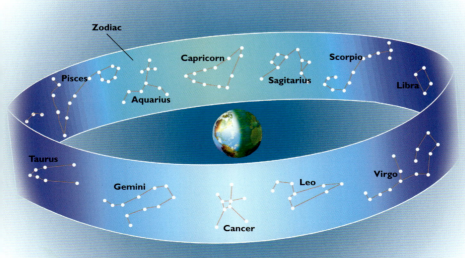

From Earth, we see the Sun and the planets travel through the same patterns of stars — the constellations of the zodiac.

Zodiac

Capricorn
Scorpio
Pisces
Sagittarius
Libra
Aquarius
Taurus
Gemini
Leo
Virgo
Cancer

Jupiter spins round incredibly quickly for such a huge body. Its day is less than 10 hours long. On the other hand, Venus spins round incredibly slowly. Its day is 243 Earth-days long.

The Earth does not spin in an upright position in relation to its path around the Sun. Its axis is tilted slightly (23⅓ degrees) in space. It is this tilt that brings about the temperature and weather changes that mark the seasons. It causes the Sun to appear higher or lower in the sky as time goes by. Summer is the season when the Sun appears highest in the sky, bringing the warmest weather. Winter is the season when the Sun appears lowest, bringing the coldest weather.

Most of the other planets have tilted axes, too. For example, Mars's axis is tilted about the same as Earth's, and the planet has seasons in the same way as Earth. Uranus has the most tilted axis. It is tilted over more than 90 degrees. This means that it spins on its side compared with the other planets.

Year by year

Earth spins round on its axis 365¼ times as it travels once in its orbit around the Sun. In other word, the journey takes Earth 365¼ days. This is Earth's orbital period, or year. All the planets have different orbital periods, or "years".

The length of a planet's year depends on where it is located in the solar system. The inner planets have much shorter years because they don't have so far to travel as they circle the Sun. Mercury has the shortest year because it is closest to the Sun. Its year is only 88 Earth-days long. Pluto has the longest year because it is farthest from the Sun. Its year is nearly 249 Earth-years long.

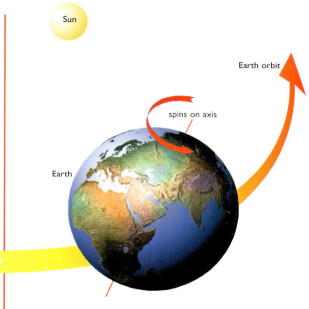

Earth spins around on its axis as it travels in orbit around the Sun.

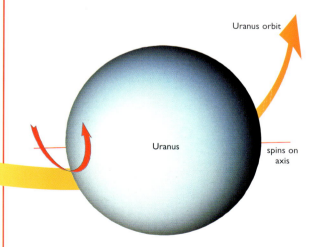

The other planets also spin on their axes as they travel. Uranus's axis is highly titlted.

Planets Large and Small

To us, Earth seems to be a big place. And it is true that it is bigger than the other three planets that are its neighbours — Mercury, Venus, and Mars. But compared with the next four planets farther out, Earth is a dwarf. These four giant planets are Jupiter, Saturn, Uranus, and Neptune.

Jupiter is staggeringly large. If Jupiter were the size of a beach ball, Earth would be only the size of a marble. Indeed, Jupiter could swallow all the other planets put together, and still have room for more. The only body in the solar system bigger than Jupiter is the Sun itself. Compared with the Sun, even Jupiter is a dwarf.

The planet we call home

Our home planet, Earth, occupies a fortunate location in the solar system, which sets it apart from all the other planets. It is not too close, nor too far away from the Sun, and so it is not too hot, nor too cold. And, at Earth's temperatures, water can exist as a liquid. The Earth is also just the right size so that its gravity, or pull, is strong enough to hold onto an atmosphere.

These three things — warmth, liquid water, and an atmosphere — make Earth a comfortable place for an amazing variety of living things. No other planet has the same conditions. So they have no plants, no animals, and as far as we know not even primitive living organisms like bacteria.

Jupiter

Uranus

Neptune

Saturn

Sun

Earth

Venus

Mars

Mercury

Pluto

The planets differ greatly in size, from tiny Pluto to gigantic Jupiter. But all of them are dwarfed by the Sun.

Also, the other planets are not as active geologically as Earth is. In other words, they do not change much inside or on the surface. But Earth is changing all the time. Its outer layer, or crust, is split into sections called plates. And these plates are constantly moving and pushing against one another. This causes the continents to drift, earthquakes to happen, volcanoes to erupt, and mountains to form.

Earth's surface is also changing all the time because of erosion. This is the wearing away and breaking down of surface rocks by the weather, flowing water, moving glaciers, and so on. We human beings are also causing changes, of course, by engineering works and mining. We are also beginning to change the atmosphere and climate of Earth, and not for the better. We really have to take better care of our planet because it is the only place we know that has life of any kind.

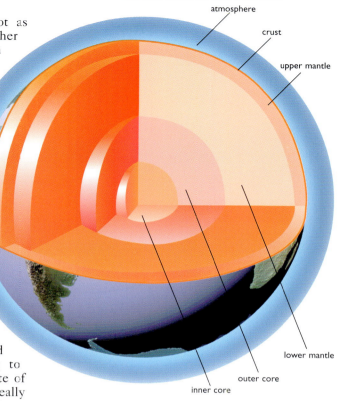

atmosphere

crust

upper mantle

lower mantle

outer core

inner core

Earth is made up of layers of rock, with a core of metal at the centre.

Planets Data

	Av. distance from Sun million miles (km)	Diameter at equator miles (km)	Circles Sun in	Mass (Earth=1)	Density (water=1)	Number of moons
Mercury	36 (58)	3,032 (4,880)	88 days	0.06	5.4	0
Venus	67 (108)	7,521 (12,104)	225 days	0.8	5.2	0
Earth	93 (150)	7,927 (12,756)	365.25 days	1	5.5	1
Mars	142 (228)	4,221 (6,792)	687 days	0.1	3.9	2
Jupiter	483 (778)	88,850 (142,980)	11.9 years	318	1.3	16
Saturn	888 (1,429)	74,900 (120,540)	29.4 years	95	0.7	18
Uranus	1,787 (2,875)	31,765 (51,120)	83.7 years	15	1.3	18
Neptune	2,799 (4,504)	30,779 (49,530)	163.7 years	17	1.6	8
Pluto	3,676 (5,916)	1,429 (2,300)	248 years	0.002	2.0	1

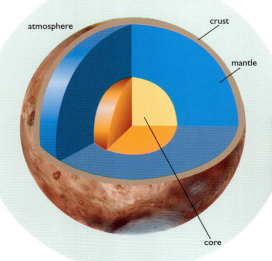

atmosphere

crust

mantle

core

Left: Like earth. Mars is a rocky planet, with a hard crust. It has only a thin atmosphere.

Below: Venus's surface is covered with lava flows from hundreds of volcanoes.

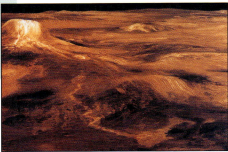

The Dwarfs and the Giants

The four planets in the inner part of the solar system are dwarfs compared with the next four planets farther out. But they differ from the outer planets in many other ways too, including their make-up, surface, atmosphere, and moons.

The other three dwarf planets that lie with Earth in the inner part of the solar system are similar to our planet in make-up. That is why they are often called the terrestrial, or Earth-like planets. They have a metal centre, or core, and layers of rock on top.

Each planet has a different kind of surface, however. Mercury's surface is almost completely covered with craters made by meteorites. It is very ancient. Venus's surface is mainly low rolling plains, covered by vast lava flows from the hundreds of volcanoes that dot the landscape. There are few meteorite craters on Venus, which tells us that its surface is young. Mars has some cratered regions and many plains. The surface is covered with rust-red soil, telling us that erosion has been at work.

Erosion on Mars has been brought about mainly by the wind, because the planet has a slight atmosphere. Venus has an atmosphere, too, but it is very much thicker, or denser. In fact, it is much denser than Earth's atmosphere. Mercury has virtually no atmosphere at all.

Between them the dwarf inner planets have only three moons. Earth has one — the Moon. And Mars has two — Phobos and Deimos — but they are both very small.

The giant gas balls

In make-up, the giant planets Jupiter, Saturn, Uranus, and Neptune could not be more different from the inner rocky planets. They have no hard surface whatsoever. They are made up mainly of gas and liquid.

Jupiter and Saturn, for example, each have an atmosphere thousands of miles deep. And the main gases in the atmosphere are hydrogen and helium. At the bottom of the atmosphere, pressures are so great that they turn the hydrogen gas into a liquid. The liquid hydrogen forms a planet-wide ocean thousands of miles deep. At the bottom of this ocean, fantastic pressures compress (squeeze) the hydrogen particles so much that they form a kind of liquid metal. Only right at the centre of the planet is there a small rocky core.

Uranus and Neptune also have deep atmospheres of hydrogen and helium above planet-wide oceans. But on these planets, the oceans seem to be made up of water, methane, and ammonia. These planets have a small core of rock and ice.

Whereas the inner planets have only three moons between them, the giant planets have many. In all, they have more than 60 moons between them. Some are only a few tens of miles across, but others are as big as planets.

And there is a final major difference between the inner planets and the giants. All the giants have rings around them. The inner planets have no rings.

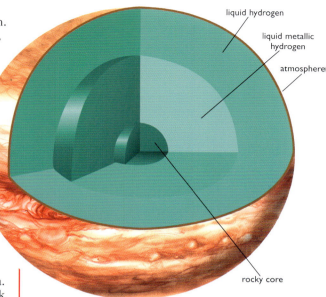

liquid hydrogen

liquid metallic hydrogen

atmosphere

rocky core

Above: Jupiter is made up of mainly of hydrogen in a liquid form, and the atmosphere is mainly hydrogen gas.

Below: Pluto's icy surface may look something like this.

Odd planet out

The ninth and outermost planet, Pluto, is not like any of the other planets. It is very tiny — only about two-thirds as big across as the Moon. It seems to be made up of a large rocky core, covered in a thick layer of ice. Astronomers think that Pluto could be the largest of a whole group of similar small bodies existing in the outer solar system, in a region known as the Kuiper Belt.

TINY WORLDS

After the planets formed billions of years ago, there were many bits of matter left over. These bits still wander around the solar system and, from time to time, make themselves known. We see some bits as meteors and comets. Others form the asteroids.

On any clear night you go stargazing, you are almost certain to spot one or more bright streaks of light in the sky. It looks as if some of the stars are falling down or travelling to another part of the heavens. That is why we often call these streaks falling or shooting stars. But their correct name is meteors.

Meteors are produced by small bits of rock or metal from outer space burning up as they travel through Earth's atmosphere. Outer space is full of these tiny bits of matter, known as meteoroids. When they get near the Earth, they fall into the clutches of Earth's pull, or gravity. Travelling at speeds of up to 45 miles (70 km) a second, they head toward the ground.

Most meteoroids are not much bigger than a grain of sand. As they plunge into Earth's atmosphere, they rub against the air molecules. This rubbing, or friction, makes them heat up. They get hotter and hotter and start to glow; then they catch fire and burn up. We see the fiery trails they leave behind as meteors. Usually, the meteoroids have burned up completely by the time they have reached a height of 50 miles (80 km) above the ground.

Most meteoroids, however, are even tinier than a sand grain, more like grains of pollen. They drift leisurely through the atmosphere and take some time to settle on the ground.

A few meteoroids, on the other hand, are really big. They create a spectacular fireworks display as they zoom through the air. Known as fireballs or bolides, they often glow yellow, red or green. They leave behind a broad and longer-lasting trail. You can sometimes hear a noise like a peal of thunder as they pass overhead. It is the sonic boom caused by them travelling at supersonic speed.

Opposite: A colour-enhanced picture of Halley's Comet, when it last returned to our skies in 1986. Telescopes were needed to see it clearly.

Left: Meteorites such as this one rain down on Earth all the time.

Showers of meteors

On most nights of the year, you may be lucky enough to spot five or six meteors every hour. They can be seen in any part of the sky, and are known as sporadic meteors.

At certain times of the year, however, many more meteors than usual may be seen. And they all come from the same part of the sky. When this happens, we call it a meteor shower.

There are quite a number of meteor showers, which occur at the same times every year. For example, in early August, swarms of meteors appear from the direction of the constellation Perseus. We call this shower the Perseids. More than 50 meteors an hour may be seen. Around mid-November, there is a shower known as the Leonids because it comes from the direction of Leo. In some years, hundreds of Leonid meteors may be seen every hour. In exceptional years, such as 1999, thousands may be seen.

Just why do meteor showers occur? The answer is because of comets. Every year, as Earth travels in its orbit around the Sun, it crosses the orbit of several regular, or periodic comets. Comets always leave a cloud of dust behind them, and every time Earth moves into this cloud a meteor shower occurs.

Above: Meteors flash through the sky during a Leonid meteor shower in mid-November.

Below: A look inside a stony meteorite.

Below Left: Iron meteorites usually also contain nickel as well.

The famous Meteor Crater in Arizona, which measures some 4,150 feet (1,265 metres) in diameter and about 575 feet (175 metres) deep.

Hitting the ground

Most of the meteoroids Earth attracts are tiny. And they burn up high in the atmosphere. But occasionally a much bigger meteoroid comes along. Its surface melts and burns away as it plunges through the air, but the rest carries on down and hits the ground. We call the lump that hits the ground a meteorite.

Most of the meteorites that fall are made up of rocky material. They are known as stony meteorites, or stones. Most of the others that fall are made up of metal, mainly iron and nickel. They are called iron meteorites, or irons. A few meteorites are made up of a mixture of rock and metal and are called stony-irons.

Thousands of meteorites have been found scattered around the world. Most are quite small, but a few are huge. The largest was found in 1920 near the town of Hoba in Namibia, in south-west Africa. It weighs about 60 tons. Like all the large specimens found, it is an iron meteorite. In general, irons are stronger than stones and are therefore less likely to break up when they hit the ground.

When meteorites hit the ground, they are travelling very fast. So they often dig out a pit, or crater. The bigger the meteorite, the bigger is the crater it digs. Fortunately, most of the meteorites that hit Earth today are small. They may dig little craters but cause little if any damage. But in the past large meteorites have hit Earth and dug huge craters that we can still see today.

The best-known crater is Meteor Crater in the Arizona Desert, which is nearly a mile across. It is very well preserved for its age of about 50,000 years. Other large craters include the ½-mile (850-metre) Wolf Creek crater in Australia. Some big craters, like the 2-mile (3.2 km) New Quebec crater in Canada, may not be a meteorite crater. It could have been formed by movements in Earth's crust.

Comets in the Sky

Meteors flash through the sky every night. But we see them only for a second or so. We may have to wait years for the greatest night-sky spectacular of all — a brilliant comet. With a huge glowing head and tails streaming out behind, the most brilliant of comets can outshine the stars. And they can hang in the sky for months.

The brightest comets, like Hale-Bopp in 1997, seem suddenly to appear in the sky. And then in a few weeks or months disappear again. This sudden appearance and disappearance used to worry ancient peoples, for to them what happened in the heavens was very important. They thought that somehow the heavenly bodies affected their lives. Comets, they thought, were evil, causing diseases and droughts, wars and destruction.

Halley and the comets

It was only in the 1500s that astronomers realized that comets must travel in outer space. Before, they thought that comets might be strange happenings in the atmosphere. In the 1600s, an English astronomer named Edmond Halley worked out that some comets return to Earth's skies again and again.

By checking the orbits of comets that appeared in 1531, 1607, and 1682, Halley decided that they must be the same body. And he predicted that it would return in 1758, after his death. The comet duly appeared, and henceforth became known as Halley's Comet. Astronomical records show that this comet has been seen about every 75/76 years since 240 BC. It last put in an appearance in 1986, and should next be sighted in 2061.

Comets like Halley, which appear regularly in Earth's skies, are called periodic comets. We know of many comets like this. The one with the shortest period — the time it takes to return — is Comet Encke, which appears every 3.3 years. Bright comets like Hale-Bopp, on the other hand, may not return for thousands of years. We call them long-period comets.

Above: Edmond Halley was the 17th century British astronomer who first suggested that comets can be regular visitors to our skies.

Below: This close-up image of Halley's comet was taken by the Giotto probe in 1986.

Dirty snowballs

What exactly is a comet — this great shining body that can stretch maybe millions of miles across the heavens? What we see is a glowing mass of gas and dust. The only solid part lies deep inside the head, or coma, of the comet. It is called the nucleus. The coma itself may be more than a million miles (1.6 million km) across, but the nucleus inside it is usually less than a mile (1.6 km) wide.

The nucleus is often described as a "dirty snowball" because it is a lump of water ice mixed with dust. When a comet nears the Sun, the Sun's warmth turns the ice into gas. This releases lots of dust as well. And the gas and dust form a great cloud around the nucleus, which we see as the coma.

Below: The comet Ikeya-Seki of 1965 was one of the brightest comets of the 20th century.

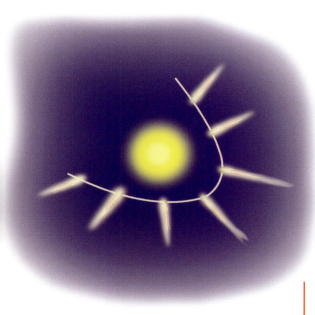

when they travel into the warmer inner part of the solar system do they release gas and dust and start to glow.

Astronomers think that billions of deep-frozen comets lie at the very edge of the solar system in a region they call the Oort Cloud. The Cloud may extend to more than one light-year (about 6 trillion miles, 10 trillion km) away from the Sun. The long-period comets probably come from this Cloud.

Closer in, there appears to be another region containing comets. It is called the Kuiper belt. The shorter-period comets probably come from this belt. Many larger bodies are found in the same region, and are known as Kuiper-belt objects. On average, they seem to be about 60-200 miles (100-300 km) across. The planet Pluto and its moon Charon orbit also lie within the Kuiper belt.

Comet tails

The Sun pours out into space all kinds of rays, or radiation, and also streams of particles that we call the solar wind. The sunlight and the solar wind push against the cloud of gas and dust of the comet. They force it away from the nucleus into a tail, or usually, two tails. The dust forms a yellowish, curving tail. The gas forms a much straighter tail that is bluish in colour.

Where comets lie

Comets circle around the Sun just like the planets. But most of them have very long orbits, which take them way beyond the planets into the very cold outer parts of the solar system. For much of their orbits, they remain deep-frozen solid lumps. And they are too tiny to be visible from Earth. Only

The Asteroids

Comets are small icy lumps left over after the planets formed and lie very far away from Earth. But there are much larger lumps of left-over matter very much closer. They are known as the asteroids, or minor planets. Most asteroids circle the Sun between the orbits of Mars and Jupiter in a broad ring, called the asteroid belt. The centre of the belt lies about 250 million miles (400 million km) from the Sun.

Only one of the asteroids, Vesta, can be seen with the naked eye. Discovered in 1807, it measures about 320 miles (510 km) across. It is the third largest asteroid, after Ceres (580 miles, 935 km) and Pallas (325 miles, 525 km). Ceres was also the first asteroid discovered, by the Italian astronomer Giuseppe Piazzi in 1801.

By studying the light reflected back by asteroids, astronomers have long known that most asteroids are irregular in shape. And so it proved with Gaspra and Ida, the first two asteroids that were photographed, by the probe Galileo.

Rock and metal

Astronomers think that the largest asteroids have a similar make-up to the inner planets, with a core of metal and rock on the outside. But asteroids must always be colliding and knocking one another to pieces.

This means that some asteroids must be now lumps of just rock, some lumps of just metal, and others a mixture or rock and metal. Astronomers have indeed found that some asteroids are made of metal. They have found this out again by studying the light they reflect.

Left: This dark meteorite probably came from the third-largest asteroid, Vesta.

Below: Most of the asteroids circle the Sun in between the orbits of Mars and jupiter, in the so-called asteroid belt.

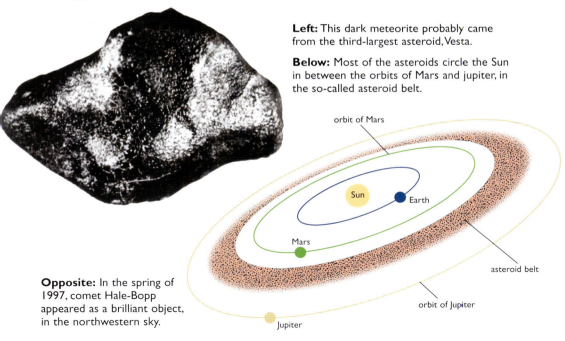

orbit of Mars

Sun

Earth

Mars

asteroid belt

orbit of Jupiter

Jupiter

Opposite: In the spring of 1997, comet Hale-Bopp appeared as a brilliant object, in the northwestern sky.

27

The asteroids that wander

We already know of more than 20,000 asteroids, some as small as 1 mile (1.6 km) across. And altogether there are probably millions. Most circle the Sun within the asteroid belt, but others wander much farther afield. Some are found in Jupiter's orbit travelling in two groups, one in front of and the other behind the giant planet. They are known as the Trojan asteroids.

Some of the asteroids have orbits that criss-cross Earth's orbit around the Sun. This means that at times these asteroids might come dangerously close to Earth. They are called NEOs, or near-Earth objects. The remains of comets are other bodies that could one day threaten Earth.

Shaping the solar system

Asteroid-like bodies, and their relatives the meteorites, have been circling among the planets ever since the solar system was born. Over the ages, they have played a major part in shaping the planets and their moons.

There were many more large chunks whizzing round the solar system billions of years ago. And they bombarded the planets and their moons without mercy. We see the result of this bombardment today in the crater-covered surface of Mercury and our own Moon. The great seas of the Moon, for example, were once great craters made by asteroids, which later filled with lava from erupting volcanoes.

Target Earth

Earth itself was heavily bombarded billions of years ago. But the craters have long since been destroyed by the upheavals that have taken place in Earth's crust since then. There is, however, plenty of evidence of more recent asteroid impacts, such as the one that happened about 65 million years ago.

At the time, so scientists reckon, an asteroid maybe as big as 10 miles (16 km) across smashed into what is now the Yucatan Peninsula in Mexico. It created havoc, digging out a huge crater and blasting vast amounts of rock and dust into the atmosphere. Thick layers of dust formed a cloud that soon covered the whole planet and stopped

Below: Many asteroids have orbits that take them a long way from the asteroid belt. Some journey out toward Saturn, whie others venture in among the inner planets.

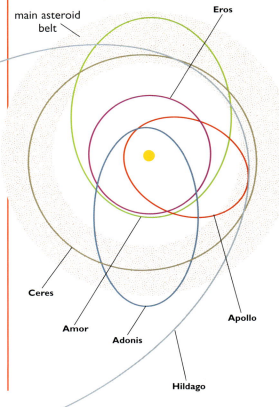

main asteroid belt

Eros

Ceres

Amor

Adonis

Apollo

Hildago

the Sun's light reaching the ground. The cloud stayed in the atmosphere for months.

Starved of light, all plants withered and died. The animals that ate plants soon starved to death. So did the meat-eating animals that usually ate the plant-eaters. Over two-thirds of all species (kinds) of living things died out, including the dinosaurs, which had ruled the Earth for more that 100 million years.

Above: The asteroid eros was first photographed by the NEAR space probe in 2000.

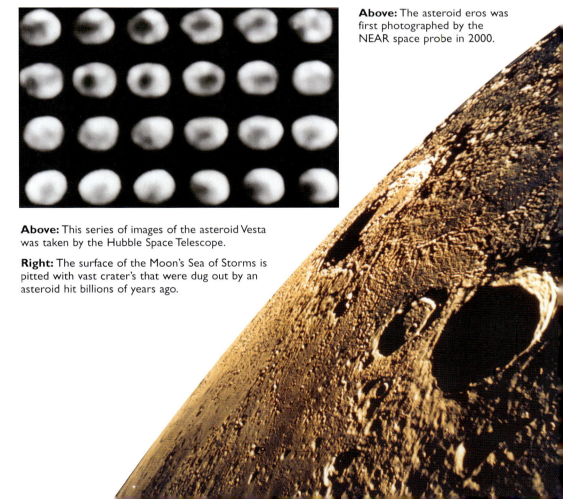

Above: This series of images of the asteroid Vesta was taken by the Hubble Space Telescope.

Right: The surface of the Moon's Sea of Storms is pitted with vast crater's that were dug out by an asteroid hit billions of years ago.

PROBING THE SOLAR SYSTEM

Space probes journey billions of miles into the depths of the solar system to encounter planets, moons, comets, and asteroids. Their cameras show us unbelievable sights that telescopes on Earth could never reveal, from giant hurricanes to erupting volcanoes.

Until the Space Age began, we had to rely on our eyes and telescopes to gather information about the heavenly bodies. In telescopes, we can see the Moon easily because it is so close. We can see some vague markings on Mars, colourful bands around Jupiter, and Saturn's shining rings. But we can see no details at all of the other planets or of any other moons except our own. The main reason is because they lie too far away.

To overcome this problem, astronomers send spacecraft to visit the planets and other heavenly bodies. These craft are called space probes. They are launched by rockets or by the space shuttle. They have to reach a colossal speed to escape from Earth's gravity — some 25,000 miles (40,000 km) an hour. This speed is called Earth's escape velocity.

Probes carry cameras, radios, and many other kinds of instruments, such as magnetometers to record magnetism and detectors to record different kinds of radiation. They work by electricity.

Probes to the near planets are fitted with solar panels to make electricity from sunlight. Probes travelling farther away from the Sun are fitted with nuclear batteries. These batteries make electricity from the heat given out by substances called radioisotopes, which give off nuclear radiation.

Probes can take years to reach their target. Their target is moving of course. So they have to be aimed, not at the target, but at a point in space where the target will be at a certain time in the future. The meeting between a probe and its target is called the encounter.

Space scientists can send their probes with incredible accuracy. For example, the Voyager 2 probe reached its last planet, Neptune, after travelling for 12 years and 4.4 billion miles (7 billion km). It skimmed just 3,000 miles (5,000 km) above the cloud tops.

Opposite: This stunning view of a massive volcano on Venus was sent back by the Magellan radar probe.

Destination Moon

The Space Age began when Russian scientists launched the first satellite, Sputnik 1, into space on October 4, 1957. US scientists launched their first satellite, Explorer 1, on January 31, 1958.

Both countries soon began trying to launch probes to the Moon. They chose the Moon as their first target in outer space because it lies very much closer to Earth than any other heavenly body. It lies only about 240,000 miles (385,000 km) away; the next nearest body, Venus, lies over 100 times farther away.

Russian scientists reached the Moon first, when their probe Luna 2 crashed-landed on the Moon in September 1959. Three weeks later, Luna 3 sent back the first photographs of the far side of the Moon, which we can never see from Earth. US scientists first

Right: The mighty Saturn V rocket boosted the Apollo astaonauts on their epic journeys to the Moon.

Below: This view of the Moon taken by Apollo astronauts shows mainly the far side, which we can never see from earth.

achieved success with Ranger 7 in July 1964, which sent back fine close-up views of the lunar surface before crash-landing.

Many probes followed. Some went into orbit around the Moon and mapped its surface. Others landed, took close-up pictures of the surface, and tested the soil. US probes gained the information needed to help NASA scientists plan the most exciting adventure of all time — the Apollo project to land astronauts on the Moon.

Walking on the Moon

To try for a Moon landing, a huge new rocket and a new three-part spacecraft had to be developed. They were tested together for the first time in December 1968, when three astronauts in Apollo 8 set off on the first human voyage to the Moon. Their mission was a triumph. They circled around the Moon and on television showed people back on planet Earth close-up pictures of the lunar surface for the first time.

Here, were the great dusty plains that ancient astronomers called seas. There, were lofty mountain ranges pierced by valleys. And everywhere there were craters. The lunarsurface was bare but oh so beautiful.

Apollo 8's journey to the Moon and back paved the way for six successful Moon landings, beginning with Apollo 11 in July 1969. On this mission, Neil Armstrong and Edwin Aldrin roamed around the surface of the Sea of Tranquillity for some 2½ hours. They set the pattern for the other landing missions, collecting rock samples, carrying out experiments, and setting up scientific equipment that would radio information back to Earth after they left.

An astronaut salutes the Stars and Stripes at the Apollo 15 landing site, in the foothills of the Apennine Mountains.

Roving far and wide

The next two landing missions, Apollo 12 and Apollo 14, also landed on sea sites — on the sprawling Ocean of Storms. The three other missions landed amid spectacular mountain scenery in the lunar highlands.

In all, the Apollo astronauts roamed around the Moon's surface for 80 hours, brought back 842 pounds (382 kg) of Moon rock and soil. They took more than 20,000 photographs, which still amaze us even today. Thanks to the Apollo missions, we now understand the Moon almost as well as we understand Earth.

An icy Moon?

Only a few probes have been sent to the Moon since the Apollo landings. But in 1994, Clementine sent back some interesting information. It detected signs of water ice in craters near the Moon's south pole.

Below: On the last three Apollo missions, the astronauts had transoprtation in the form of the lunar rover, or Moon buggy.

A follow-up mission by Lunar Prospector four years later confirmed that ice does seem to be present, not only near the south pole but near the north pole as well. Astronomers think that the ice would have come from comets that crashed on the Moon long ago.

One day, astronauts may return to the Moon and set up bases there. If there is ice, they will be able to use it for their water supply instead of bringing water from Earth. The water could also be used to make fuel for their rockets.

Above: This typical example of Moon rock is made up of bits of volcanic rock stuck together.

Left: Mariner 10 took the first close-up pictures of Mercury.

Visiting Neighbours

After the Moon, the planet Venus is our closest neighbour in space. Mariner 2 was the first successful probe to fly near Venus in 1962. It reported that the planet had a temperature of hundreds of degrees. It also found that carbon dioxide was the main gas in Venus's atmosphere. Later, Russian Venera probes parachuted through Venus's clouds. They confirmed the planet's high temperature and thick carbon dioxide atmosphere.

Peering through the clouds

In 1974, Mariner 10 sent back photographs of Venus on its way to Mercury. They showed bands of clouds in the atmosphere but couldn't see down to the surface. The clouds were too thick.

Astronomers on Earth, however, were starting to map Venus's surface using radar. They used radio telescopes as radar dishes to bounce radio waves from Venus's surface. Unlike light rays, radio waves can "see" through clouds.

Since then, all of Venus has been mapped by probes that have gone into orbit around the planet and scanned its surface by radar. The most successful was Magellan, which studied the planet between 1990 and 1994. It sent back detailed images that showed a landscape of towering volcanoes, vast lava plains, and deep, snaking valleys.

Lunar landscapes

When Mariner 10 reached Mercury in March 1974, it discovered that the planet is almost completely covered in craters. It looks much like some parts of the Moon. But, unlike the Moon, Mercury does not have any large seas, or plains. Its largest feature is the huge Caloris Basin, 830 miles (1,340 km) across. The Basin was created billions of years ago when a huge asteroid smashed into the planet.

Below: Magellan used radar to peer through venus's clouds.

Mariners to Mars

Mariner spacecraft also pioneered the robot exploration of Mars, beginning in 1965 with Mariner 4. This was the first probe ever to send back pictures of another planet. They showed a generally cratered landscape. Later probes, however, showed that the Martian landscape varies widely from region to region.

In 1971, Mariner 9 went into orbit around Mars and eventually mapped most of the planet. It discovered huge volcanoes, higher than Earth's highest mountain, Mount Everest. It spotted a huge canyon system far bigger than the Grand Canyon in Arizona. The canyon was named Valles Marineris, or Mariner Valley.

The Viking landings

The exploration of Mars continued when two identical probes, Viking 1 and Viking 2, went into orbit around Mars in 1976. Later they dropped landing craft onto the Martian plains Chryse and Utopia. The orbiters remained in orbit taking photographs of the whole planet.

The pictures the landers sent back showed a remarkably similar landscape at both landing sites. Loose soil covered the surface, and small rocks were strewn around everywhere. Both soil and rocks were a rusty red colour.

Among other equipment, the landers carried weather instruments and a small laboratory. In the laboratory, they tested samples of soil to see if they could find traces of Martian life. But they found nothing.

Wheels on Mars

No probes returned to the Red Planet until July 1997, when Pathfinder landed. The landing site was in a region that astronomers think was once the mouth of a river system. Pathfinder's first pictures showed a landscape that looked very much like that at the two Viking landing sites. Rust-red rocks lay scattered about in the rust-red soil.

A viking lander uses its retrorockets to slow it down for a soft landing on the Red Planet Mars.

Pathfinder also carried a little four-wheeled rover, named Sojourner. Its job was to visit some of the rocks and find out what kind they were. Sojourner found that there were several different rock types. This is what would be expected in a river mouth, where they would have been washed down from different places upstream.

By the fall of 1997, Pathfinder had sent its last communications, but another spacecraft was just beginning its mission. It was named Mars Global Surveyor. It sent back highly detailed pictures of the Martian surface, showing a host of features not spotted before. Some showed clear signs that water may once have flowed on Mars.

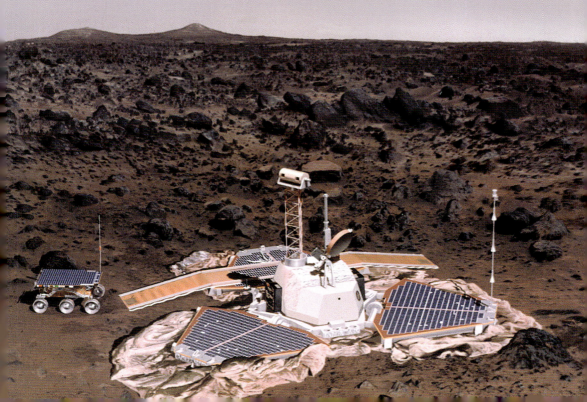

Pathfinder set down on Mars. On the left is the little Sojourner rover, before it set out to explore the nearby rocks. In the distance on the horizon are the mountains known as Twin Peaks.

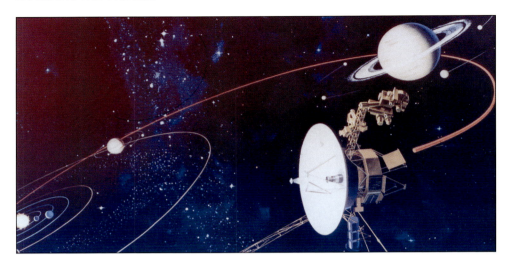

To Distant Worlds

The outer planets lie very much farther away from us than our neighbours Venus and Mars. Jupiter never comes closer than about 400 million miles (600 million km). Saturn never comes within 800 million miles (1,200 million km). And the other outer planets are billions of miles further away still.

Guiding a tiny probe to a target so far away is incredibly difficult. If it is going slightly too fast or slightly too slow, or starts off in slightly the wrong direction, it will miss its target by millions of miles.

Above: The Voyager probes to the outer planets were of identical design. They radioed back pictures and data from their 12-foot (3.7-metre) dish antenna.

However, Pioneer 10, the very first probe launched to Jupiter in March 1972, made its encounter with the planet exactly on time. In December 1973, it took the first close-up pictures of Jupiter, showing that the Great Red Spot is a gigantic storm.

A year later an identical craft, Pioneer 11, made another successful encounter with Jupiter. Afterward it set course for another target, Saturn, and was renamed Pioneer-Saturn. After a five-year journey, it reached the ringed planet in 1979. It sent back the best pictures yet of Saturn and its beautiful rings. It also discovered a new ring and a new moon.

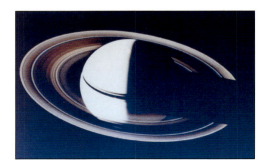

Left: Voyager 2 looked back at Saturn as it began its five-year journey to its next target — the planet Uranus.

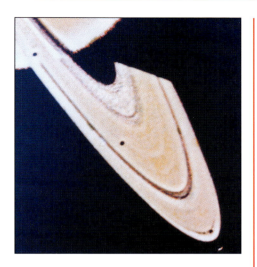

Above: The Voyagers also took pictures of many of the giant planets' moons. This shapeless lump is saturn's moon Hyperion, about 225 miles (360 km) long.

Left: Pioneer-Saturn took the first close-up pictures of the ringed planet Saturn.

The Voyager missions

After their encounters with Jupiter, the Voyager probes used the gravity of the giant planet to sling them into a new and faster trajectory (path) toward Saturn. Voyager 1 streaked past Saturn in November 1980. Voyager 2 did not make a close approach until August 1981.

Both Voyagers performed magnificently, showing the planet as never before. They spotted new moons and new rings, and showed that the three broad rings we can see from Earth are actually made up of thousands of separate ringlets.

To other worlds

After its Saturn encounter, Voyager 1's mission was over. But Voyager 2 had other worlds to conquer. The longer route it had taken from Jupiter to Saturn had put it on course for a visit to Uranus in 1986 and Neptune in 1989. It was taking advantage of a rare lining-up of the giant planets that would not occur again for 175 years.

Below: Voyager I looked into the heart of Jupiter's Great Red Spot. This furious super-hurricane, hundreds of times bigger than those on Earth, has been raging for centuries.

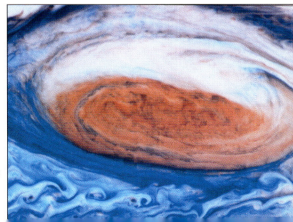

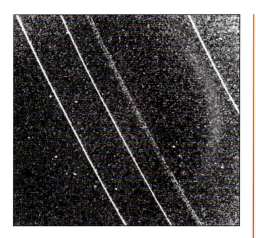

Shown here are 4 of the 11 rings Voyager 2 spotted while circling Uranus. They had first been detected by astronomers on Earth.

Voyager 2 provided this close-up of Neptune's Great Dark Spot, which is a huge storm region, edged with white clouds.

Encounters with Uranus and Neptune

Voyager 2 flew past Uranus in January 1986, revealing the planet and its rings clearly for the first time. It spotted no less than 10 new moons. Of the known moons, Miranda proved to be the most interesting, with the most amazing patchwork surface.

Then on to Neptune. Voyager 2 encountered the planet in August 1989. Pictures showed that Neptune's atmosphere is a deeper blue-green colour than that of Uranus and has much more weather. There are patches of white clouds and dark spots, which are violent storms. As on all its previous encounters, Voyager 2 found several new moons. And it also spotted a system of rings. This meant that all four giant planets had rings.

Voyager 2's final encounter was with Neptune's largest moon, Triton. This proved to be a fascinating body with a strange surface. In places it was covered with pink snow, and icy volcanoes were erupting.

Having encountered four planets in 12 years, Voyager 2 proved to be the outstanding probe of the Space Age. It is still working today, sending back information about interplanetary space. Its batteries should keep it working until about the year 2020. By then it may have left the solar system and be on its way to the stars.

Galileo and Cassini

Study of the outer planets from space continued in 1990. But it was a much more distant study — from the Hubble Space Telescope. By then another probe was starting its journey to Jupiter. It was named Galileo. It followed a roundabout route that took it twice round the Sun and past Venus and Earth. It travelled twice through the asteroid belt, where it took the first close-up pictures of asteroids Gaspra and Ida.

Galileo did not reach Jupiter until December 1995. Before it went into orbit around the giant planet, it dropped a small probe into the atmosphere tooreport on conditions there. For more than four years, Galileo followed the changes taking place in Jupiter's atmosphere and on the surface of its large moons.

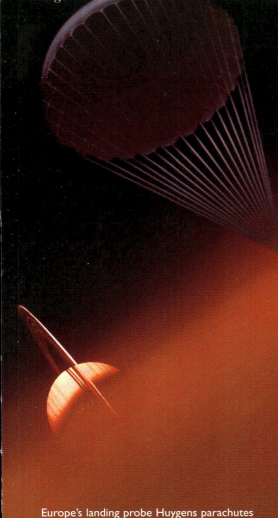

As Galileo's mission was drawing to a close, another probe was winging its way to Saturn. Launched in October 1997, it was named Cassini-Huygens. The mission plan called for the probe to reach Saturn in July 2004. The main part of the probe, Cassini, is designed to orbit Saturn for at least four years to study its atmosphere, rings, and moons.

Huygens is a smaller probe, designed to land on Titan, Saturn's largest moon. Titan is unique among moons because it has a thick atmosphere. Scientists are dying to know what lies underneath it. There could be a strange landscape with rivers of liquid methane and drifts of methane snow.

Catching comets

In 2004 too, in another part of the solar system, a probe called Stardust headed for home after a rendezvous with Comet Wild 2. It carryied dust from the comet, and delivered it to Earth early in 2006.

Mission scientists are keen to analyse comet particles, which will help them understand better how the solar system began and developed. They think they might even find simple carbon compounds like those found in living things. This could suggest that comets might be carriers of life in space.

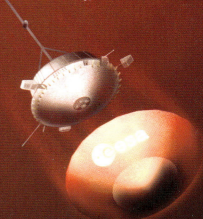

Europe's landing probe Huygens parachutes through the thick orange atmosphere of Titan, Saturn's biggest moon.

Space is full of dazzling clouds of gas like this, which astronomers call nebulars. These clouds are the birthplace of stars.

Exploring the Stars and Galaxies

On every clear night, the twinkling stars shine out of the velvety blackness of space, pouring their faint light onto the darkened Earth. You can see thousands of stars just with your eyes alone. You can see many thousands more when you look at the night sky through binoculars or a telescope.

Telescopes will also show you great bright and dark clouds of gas and dust, which astronomers call nebulas. It is in such clouds that the stars are born. Here and there, telescopes spot stars gathered together in their hundreds of thousands to form dense clusters.

The most powerful telescopes will pick out stars and nebulas and clusters in distant star islands far beyond the stars we see in our own sky. Countless billions of star islands like these, which astronomers call galaxies, make up our universe.

We explore our great universe of stars and galaxies. We discover what stars are like and how they live and die. We look at our home galaxy, to which the Sun and all the other stars in the sky belong, and at other galaxies that shine like beacons in the universe.

Finally, we look at the universe as a whole, finding that it is expanding, as if from a mighty explosion long ago. Astronomers think that a mighty explosion, or Big Bang, did actually happen. They think they know how the universe we know today came about. But they haven't yet solved the final mystery: how will the universe end?

DISTANT SUNS

Located unbelievable distances away in space, stars are found in enormous variety. Some shine feebly, others brilliantly; some are tiny dwarfs, others massive giants; some journey through space alone, others cluster together in their thousands.

To our eyes, the stars look like little points of light. Even in powerful telescopes, which magnify things thousands of times, the stars still appear as points. But this doesn't mean that they are small. What it means is that they are very, very far away from us.

If you could travel through space for many trillions of miles, you would find that the stars are great globes of hot gas that pour out light and heat into space. They are just like the globe of hot gas that appears in our skies every day — the globe we call the Sun. The stars are very distant suns.

Just how far away are the stars? The answer is: farther than we can ever imagine. Sirius, the brightest star in the sky, is one of the nearest stars to us. But astronomers tell us it is 52 trillion (52,000,000,000,000) miles or 84 trillion kilometres away.

No one can imagine such large figures. The mile or the kilometre is too tiny a unit to measure the vast distances in space.

So astronomers look at distances to the stars in another way — in terms of the time it takes their light to travel to Earth. In a year, light travels nearly 6 trillion miles (10 trillion km).

The light from Sirius takes over 8 years to reach Earth. So astronomers say that the star lies over 8 light-years away. They are using the light-year as a unit to measure distance. Using this unit makes distances in space much easier to understand.

Left: Many stars puff off their outer layers of gas as they die, leaving a tiny white dwarf star behind.

Opposite: Huge globes of stars are found circling around the centre of our galaxy. Called globular clusters, they can contain a million or more stars.

Magnitudes of Brightness

When you look at the night sky, you notice that the stars vary widely in brightness. Some stars stand out like beacons, while others you can hardly see.

Astronomers describe the brightness of a star by its magnitude. A Greek astronomer named Hipparchus first used this method over 2000 years ago. He said that the brightest stars you could see had a brightness of the first magnitude. The dimmest ones had a brightness of the sixth magnitude. Other stars had brightnesses in between.

Bright and dim

Astronomers still use this idea today. But they have extended the scale. They give very bright stars magnitudes smaller than 1 into negative numbers. The brighter the star, the lower the number. The brightest star we see in the sky, Sirius, has a magnitude of −1.45.

Dim stars are given a magnitude greater than 6. And the dimmer, or fainter the star, the higher the number. A faint red star in the constellation Centaurus (Centaur) has a magnitude of about 11. This star, named Proxima Centauri, is famous as the nearest star in the heavens. It lies at a distance of about 4.2 light-years.

True brightness

However, we must remember that how bright a star appears in the sky does not describe how bright it really is. This is because the stars all lie different distances away. So a truly dim star nearby may look brighter than a truly bright star a long way off.

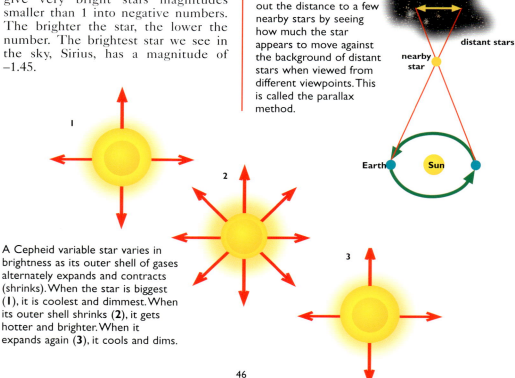

Astronomers can figure out the distance to a few nearby stars by seeing how much the star appears to move against the background of distant stars when viewed from different viewpoints. This is called the parallax method.

distant stars

nearby star

Earth

Sun

A Cepheid variable star varies in brightness as its outer shell of gases alternately expands and contracts (shrinks). When the star is biggest (1), it is coolest and dimmest. When its outer shell shrinks (2), it gets hotter and brighter. When it expands again (3), it cools and dims.

Variable stars

When we look at the stars, they twinkle. It is as though they are flickering like a candle flame and changing in brightness. But most stars shine steadily all the time. They appear to twinkle because of air currents.

However, some stars really do change in brightness. We call them variable stars because their brightness varies over time. Some variables change in brightness because they're actually a pair of stars that circle around each other. As each star regularly passes in front of the other it blocks the other's light. When this happens, the system becomes fainter. The best-known example of this kind of variable is Algol, in the constellation Perseus. It is often called the Winking Demon.

Some stars change in brightness because they change size. These include stars called Cepheids. As they expand, they become dimmer. When

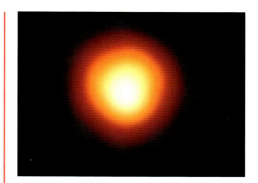

Hubble Space Telescope view of the supergiant star Betelgeuse, in Orion.

they shrink again, they become brighter. American astronomer Henrietta Leavitt pioneered work on the Cepheids early in the 20th century.

Other stars actually explode, and for a while they become thousands or even millions of times brighter than before. Then they slowly fade away.

Name	Constellation	Apparent magnitude	Distance (light years)
Sirius	Canis Major	−1.45	8.8
Canopus	Carina	−0.73	196
Alpha Centauri	Centaurus	−0.01	4.3
Arcturus	Bootes	−0.06	37
Vega	Lyra	0.04	26
Capella	Auriga	0.08	46
Rigel	Orion	0.11	815
Procyon	Canis Minor	0.35	11.4
Achernar	Eridanus	0.48	127
Beta Centauri	Centaurus	0.60	390
Altair	Aquila	0.77	16
Betelgeuse	Orion	0.80	650
Aldebaran	Taurus	0.85	68
Acrux	Crux	0.9	260
Spica	Virgo	0.96	260
Antares	Scorpius	1.00	425

What Stars Are Like

Stars are great balls of hot gas, with much the same make-up as the Sun. They pour out fantastic amounts of energy into space as light, heat, and other forms of rays.

Stars are made up mainly of two gases — hydrogen and helium. But they contain many other chemical elements as well, such as calcium, iron, and magnesium.

In size, stars vary enormously. With a diameter of 865,000 miles (1,400,000 km), the Sun is a fairly small star. It is classed as a dwarf. Yet other stars can be hundreds of times smaller.

On the other hand, many stars are very much bigger and are termed giants. They can be up to 100 times bigger across than the Sun. But even giants are tiny compared with the biggest stars of all, named supergiants. These stars can be up to 1000 times bigger across than the Sun.

On the move

The stars look as though they are fixed in the night sky. The patterns of stars, the constellations, do not appear to change from century to century. Yet the stars do move — at speeds of up to hundreds of miles a second. Some are hurtling towards us, others are rushing

The Sun is a dwarf compared with some other stars. But one day it will expand to become a red giant.

Red Giant

Blue Giant

Super Giant

Sun

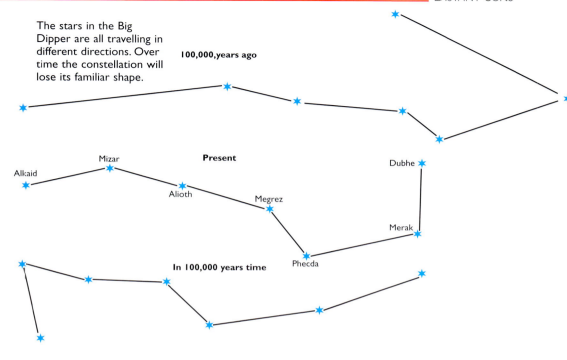

The stars in the Big Dipper are all travelling in different directions. Over time the constellation will lose its familiar shape.

100,000,years ago

Present

Mizar

Alkaid

Alioth

Megrez

Dubhe

Merak

In 100,000 years time

Phecda

away. The reason why we cannot see them move is because they are so far away.

Only with telescopes can astronomers detect any movement, and then it is only in a few hundred of the nearest stars.

Inside stars

Stars are like great fiery furnaces, pouring out energy into space. Their surface is scorching hot. The temperature of the Sun's surface is about 10,000°F (5,500°C). But the hottest stars are more than ten times hotter.

The temperatures inside stars are hard to imagine. They rise in the centre to 27,000,000°F (15,000,000°C) or more. And the pressure inside stars

is millions of times more than the air pressure on Earth.

Under these incredibly high temperatures and pressures, atoms of hydrogen gas are forced to join together. This process is called nuclear fusion. It produces fantastic amounts of energy. And it is this energy that keeps the star shining. The same process is used to produce energy in the hydrogen bomb.

When hydrogen fuses together in nuclear reactions, a new element forms — helium. This explains why helium is the main substance found in stars after hydrogen.

Stars like the Sun use up about 600 million tons of hydrogen "fuel" every second. Yet they are so big that they can keep shining for billions of years. When their hydrogen runs out, they start to die.

The white light from the Sun and the stars is actually a mixture of the colours we see displayed in the rainbow.

Looking At Starlight

The stars lie trillions of miles away. Yet astronomers can tell us how hot they are, what they are made of, how fast they are moving, and many more things besides. They gain all this information from the feeble light the stars give out.

They can tell the temperature of the surface of a star from the colour of the starlight. The coolest stars are a dull red, hotter ones bright red, and even hotter ones yellow. You can see the same kind of colour change as the heating element of an electric stove heats up. It changes colour from dull red to yellow as it gets hotter.

The hottest stars of all shine with a bluish-white light.

The colours of the rainbow

To understand how astronomers learn other things about the stars, we must first look at rainbows.

We often see a rainbow when the sun comes out following a rain shower. It is made up of different colours, from violet to red. This happens because sunlight is actually made up of different colours. They make white when they

To our eyes, all stars look much the same, shining down on us with a white or yellowish light. But in photographs stars reveal their true colours.

50

Metal changes colour as it is heated. It changes from dull to bright red, then orange, yellow, and white as the temperature goes up. We can estimate the temperature from its colour.

Left: When sunlight or starlight is passed through a prism, a spectrum of colours is produced.

Below: We can tell a lot about a star from the dark lines in its spectrum. Cool stars (bottom) have more lines in their spectrum than hot stars.

are mixed together. On a drizzly day, sunlight passes through raindrops. They split the light into its different colours, which we see as the rainbow.

We can also split sunlight into its different colours with a wedge of glass, or prism. We call this division of light into colour a spectrum. Instruments for dividing light into a spectrum are called spectroscopes.

Tell-tale lines

Astronomers use a spectroscope to get a spectrum from starlight. When they look at a star's spectrum closely, they find that it is crossed by a number of dark lines. These spectral lines provide astronomers with much information.

They can find out about its make-up. This is because certain sets of lines in the spectrm shows that the star contains certain chemical elements.

The position of the spectral lines can show the direction a star is moving. If the star is travelling toward us, all the lines move toward the blue end of the spectrum. This is called a blueshift. If the star is travelling away from us, the lines move toward the red end, which is called a redshift.

The same thing happens with sound waves, for example, when an ambulance flashes by. As it races toward us, its siren has a high note (like a blueshift). When it races away, its siren changes to a low note (like a redshift).

Space shuttle astronauts took this picture of the brightest stars in the constellation Orion. These stars look as if they are grouped together in space. But in fact they all lie at different distances from us.

Travelling Companions

One of the delights of stargazing is looking at double stars — stars that appear close together in the sky. You can see some with the naked eye, including one in the Big Dipper, part of the constellation of Ursa Major (Great Bear). The second star along the handle of the Dipper is Mizar. Close by is a fainter star called Alcor. Both can be clearly seen on a dark night as a pair, or double star.

It looks as if the two stars are close together in space, but they aren't. Alchor is much farther away than Mizar. They appear together in the sky only because they happen to lie in the same direction in space.

A closer look

When you look at Mizar in a small telescope, you find that it is not one star but two. And this time the two stars really are close together in space. They form a two-star system we call a binary.

Astronomers know of many thousands of binary stars. Sirius and Alpha

Eclipsing binaries

In some binaries, the two stars circle round each other like this, as we see them from Earth.

As the stars circle each other, each one regularly passes in front of, then behind, the other. Each time this happens, the brightness of the star system drops. This kind of system is called an eclipsing binary. The best-known example is the star Algol, in the constellation Perseus. Algol dims to one-third of its normal brightness about every 2½ days. This dimming is noticeable to the naked eye, and lasts for a few hours

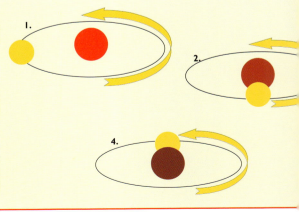

Centauri, two of the brightest stars in the sky, are both binaries.

In some binary systems, the two stars are so close together that we can never see them separately through a telescope. But we can still detect them using a spectroscope.

The two stars in the Mizar binary system each turn out to be binaries when viewed through a spectroscope. This makes Mizar a system of four stars.

Grouping together

Most stars travel through space with companions. Out of every 100 stars, only about 30 stars travel through space alone, like our Sun does.

Some stars travel together in large groups, which we call clusters. In open clusters, the stars are quite widely scattered. We can see one clearly in the Northern Hemisphere, in the constellation Taurus (Bull). It is the Pleiades. It is also called the Seven Sisters for the seven daughters of Atlas, a hero of ancient Greek mythology. Most people, however, can see only six of the seven main stars in the cluster.

Above: The stars in this fine open cluster sparkle like jewels. It contains several hundred stars altogether.

Below: In a simple binary star system (top), the two stars circle around one another. In this multiple star system (bottom), pairs of stars circle around each other.

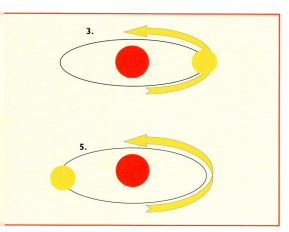

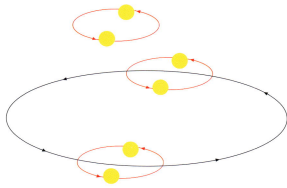

A Hubble Space Telescope view of a globular cluster in a neighbouring galaxy, the Large Magellanic Cloud. Tens of thousands of stars are on view.

Open clusters

Binoculars and telescopes show that the Pleiades contains many more stars than seven. In all, there are probably as many as 300. They are quite young stars, only about 50 million years old. They are hot and blue-white in colour.

There is also another open star cluster in Taurus, around the reddish star Aldebaran, which marks the eye of the bull. It is a V-shaped group called the Hyades.

Globular clusters

In another kind of cluster, however, the stars are packed close together into a globe shape. That is why it is called a globular cluster. In general, the stars in globular clusters are old. In fact they are some of the oldest stars in our star system, or galaxy. Studying them tells us a lot about what our galaxy was like long ago.

In the Northern Hemisphere, there is a fine globular cluster in the constellation Hercules. Known as M13, it contains hundreds of thousands of stars and is just visible to the naked eye as a faint star.

However, the finest globular clusters are to be seen in far southern skies, in the constellations Tucana (Toucan) and Centaurus (Centaur). The two clusters, 47 Tucanae and Omega Centauri, are clearly visible to the naked eye and look magnificent in binoculars.

This colourful cloud of glowing gas is the famous Orion Nebula. It is clearly visible to the naked eye.

Clouds in Space

The space between the stars is not quite empty. Particles of gas and minute specks of dust are found scattered everywhere in tiny amounts. We call this stuff interstellar matter.

In places, this matter gathers together to form denser (thicker) clouds, called nebulas (Latin for clouds). Typically, these clouds are about 30 light-years across, but others are much larger, spreading over 100 light-years or more.

Glowing bright

You can see one of the nearest nebulas in the constellation Orion. You see it as a bright patch below the three bright stars that form Orion's Belt.

The Orion Nebula is an example of a bright nebula, which shines. It shines for two reasons. One, its dust reflects the light from nearby stars. Two, its gas particles glow as they give off energy they have taken in from nearby stars.

In the dark

Many nebulas, however, do not shine. We call these dark nebulas. We can see them only when they block the light from stars or bright nebulas behind

The Trifid Nebula is one of many found in the constellation Sagittarius (Archer).

them. It is the dust in the nebulas that blocks the light. A famous example of a dark nebula is the Horsehead Nebula, also in Orion. It is well-named because it really does look like the head of a horse.

Below: A dark nebula (left) looks dark because it blots out the light of distant stars. Bright nebulas may shine because they reflect light (centre) or because their gas glows (far right).

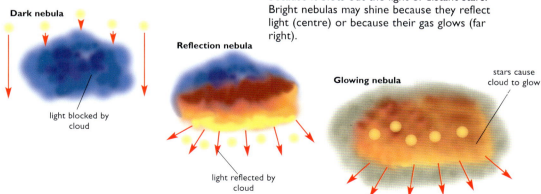

Dark nebula
light blocked by cloud

Reflection nebula
light reflected by cloud

Glowing nebula
stars cause cloud to glow

MATTERS OF LIFE AND DEATH

The stars seem to be everlasting. We see the same ones in the sky today that our ancestors saw thousands of years ago. But stars do change — over periods of millions and billions of years. They are born, they grow up, and they die, just like living things.

Astronomers can never follow all the stages in the life of a single star. This is because stars change only over a very long time — billions of years. But astronomers can study different stars at different stages of their lives. And they can then piece together how a typical star is born, lives, and dies.

In a similar way, we could learn a lot about human lives by studying the people in, say, a shopping mall. We would see babies, toddlers, teenagers, adults, and senior citizens. And over time we would probably figure out how humans live — from babies when they're young to seniors when they're old.

All stars begin their life in a huge cloud of gas and dust. There are many such clouds in the universe. They are made up mainly of hydrogen, with a sprinkling of other chemical elements.

Stars begin to form when parts of the cloud get denser, or thicker. When they get dense enough, the pull of gravity makes its particles begin to attract one another. The cloud begins to collapse, or shrink. As it gets denser, its gravity increases, and it pulls in more and more matter. As gas and dust particles rain down on the shrinking mass, they give up energy. This makes the mass heat up.

By now the mass is spinning around and has formed into a flattened ball shape. The more it shrinks, the faster it spins. This is similar to what happens to ice skaters spinning on the spot. With arms outstretched, they spin slowly, but when they draw in their arms (and shrink), they start to spin faster.

The densest part of the spinning mass is in the middle, and it is this part that may eventually become a star. The remaining matter forms into a thick disc around it.

Opposite: Big stars give off great masses of gas as they approach old age. Eventually, they may blast themselves to pieces.

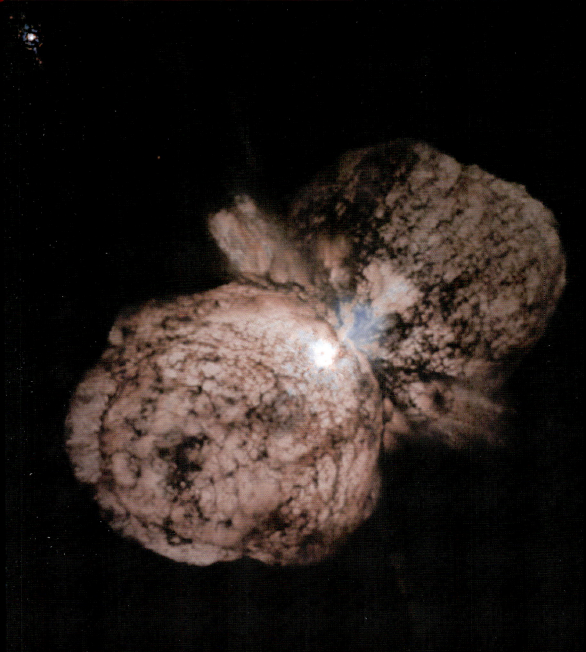

A Star is Born

Over time, the middle part of the spinning mass of matter gets hotter and hotter and starts to glow. Pressures in the centre rise higher and higher.

A dramatic change takes place when the temperature reaches about 20 million degrees F (10 million degrees C). Nuclear reactions begin: hydrogen atoms smash into one another and fuse together. Enormous energy is produced, which pours out into space as light, heat, and other radiation. The shining body is a new star.

For a while, the newborn star continues shrinking. But soon the pressure of the energy pouring outward from the core balances the pressure of the matter pulled inward by gravity. The star reaches a steady state, staying the same size and having the same energy output, or brightness.

The lives of the stars

How long a star shines steadily like this depends mainly on its mass. The smaller the mass a star has, the longer it will live.

The smallest stars have a mass only about one-tenth of the Sun's mass. They are much cooler than the Sun and give off reddish light. That is why we call them red dwarfs. These bodies live the longest of all stars. Some can probably live for as long as 200 billion years. This is more than ten times as old as the universe is now!

Like the Sun

Our own star, the Sun, has been shining steadily for about 5 billion years. Astronomers calculate that it will probably carry on shining steadily for another 5 billion years. Then, it will start to die. Stars with a similar mass to the Sun will probably shine steadily for about the same length of time before they die.

Below: Stages in the birth of a star from a nebula (**1**). The nebula shrinks and its centre gets hot (**2**). Excess gas is blown away (**3**), leaving a disk with a hot body in the centre (**4**). This body becomes a star (**5**). Surrounding matter may become planets or be blown away to leave a star that shines steadily (**6**).

Brown dwarfs

Not all the masses of matter that form out of the gas and dust clouds become stars. Some don't grow big and heavy enough. And this means that the temperatures and pressures inside them can't rise high enough to trigger off nuclear reactions. These failed stars remain warm, glowing balls, known as brown dwarfs.

Astronomers think that brown dwarfs may be similar in make-up to the planet Jupiter, but with up to 100 times more mass. They probably have a thick atmosphere of hydrogen above a deep liquid hydrogen ocean. The Hubble Space Telescope has spotted many objects that could be brown dwarfs.

Sun-like Stars

A star with a similar mass to the Sun shines steadily for about 10 billion years. During this time, it is "burning" up the hydrogen fuel in its centre, or core, in nuclear reactions to make helium.

When all the hydrogen is used up, the reactions stop. Gravity causes the core to contract, or shrink. As it shrinks, it heats up. The energy it gives off makes the outer part of the star swell up, or expand.

Over time, the star can grow up to 100 times bigger across, turning into a red giant. You can get the picture if you imagine an apple seed growing into something the size of a watermelon. When the Sun becomes a red giant, its outer layers will expand out beyond the planet Mercury, and Earth will become hotter than an oven.

The core gets hotter and hotter as it shrinks. Eventually it becomes so hot that it triggers off more nuclear reactions, this time between the helium atoms. In these reactions, the helium changes into carbon. When the helium is used up, the reactions end. The core and the rest of the star begin to contract with the force of gravity.

Smoke rings in space

As the dying star shrinks, it regularly puffs off clouds of gas. They form a kind of shell around the star, which expands outward as time goes by. The shell of gas is lit up by the energy the shrinking star gives out.

How a star like the Sun dies. After shining steadily for ages (1), it starts to expand (2, 3, 4) until it becomes a red giant (5). Over time the red giant shrinks, puffing off gas as it does so (6). All that remains is a tiny white dwarf star (7).

The core of the star grows smaller and smaller, growing ever brighter. It turns into a star we call a white dwarf.

We can see many dying stars like this in the heavens, surrounded by a glowing shell of gas. Early astronomers called them planetary nebulas because in telescopes they appear as a disc, like the planets. But planetary nebulas have nothing to do with the real planets.

A classic planetary nebula is the Ring Nebula in the constellation Lyra (Lyre). It looks like a colourful smoke ring.

Heavy dwarf

White dwarfs are white and hot. But they do not look very bright to us because they are so small. Most are only about the size of Earth. But they can contain as much mass as the Sun. So they are very heavy for their size. A cupful of their matter would probably weigh as much as 100 tons on Earth.

One of the best-known white dwarfs is a companion of Sirius, brightest star in the sky. It was also the first white dwarf to be discovered. Sirius is called the Dog Star, so its companion is often called the Pup.

The end

Over time, white dwarfs gradually cool down as their energy disperses into space. They will become dimmer and dimmer. One day they will darken and turn into black dwarfs and disappear from sight. No one knows how long this takes. Maybe the universe is not old enough for black dwarfs to have formed yet.

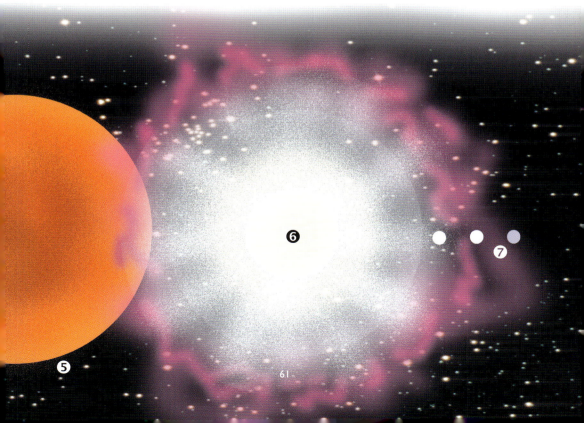

The Death of Massive Stars

Some stars are much bigger than the Sun, more than ten times the Sun's mass. These heavyweight stars burn their hydrogen fuel quickly. And they shine thousands of times more brilliantly than the Sun. But they have a much shorter life. They may shine steadily for only a few million years old before they use up their hydrogen and start to die.

As in smaller stars, the core of a big star begins to collapse and heat up when it starts to die. And its outer layers expand. In time it grows into a huge supergiant star many hundreds of times wider across than the Sun.

The temperature of the massive core of the supergiant rises to hundreds of millions of degrees. And it becomes hot enough for many other kinds of nuclear reactions to take place, which produce other chemical elements. When no

Left: Wisps of glowing gas from an ancient supernova explosion. They travel through space for many light years.

more reactions can take place, the core collapses under its own weight — in just a few seconds.

Going supernova

The energy released by the collapsing core sets off a fantastic explosion, called a supernova. The star blasts itself apart, flaring up to become billions of times brighter than it was. For a while it may become as bright as an entire galaxy of stars.

Over the past 1,000 years astronomers have recorded four supernovas in our galaxy. Chinese astronomers spotted one in 1054. They called it a ghost star. Skilled observers, they noted where it was in the heavens and that it was so bright that it could be seen in the daytime.

We see the remains of the 1054 supernova today as a cloud of gas in the constellation Taurus (Bull). We call it the Crab Nebula.

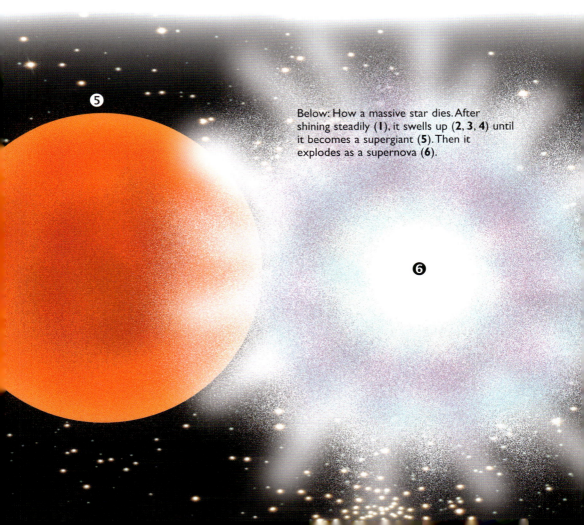

Below: How a massive star dies. After shining steadily (1), it swells up (2, 3, 4) until it becomes a supergiant (5). Then it explodes as a supernova (6).

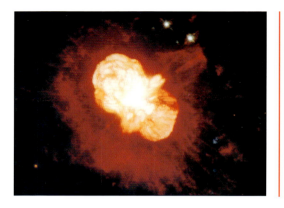

explosion — slowly expanding rings of glowing gas.

Seeding the universe

Supernovas scatter into space all the chemical elements that made up the original star. These elements — nature's building blocks — become part of the great clouds, or nebulas, in which new stars will be born. The Sun and the planets were born in such a cloud. They contain atoms and elements made in the nuclear furnace of a long-dead star. So do we.

Above: An expanding shell of gas hides one of the biggest stars known, named Eta Carinae. This star will probably not survive for long before it becomes a supernova.

Left: Rings of glowing gas surround what is left of the star that became a supernova in 1987.

Below: These delicate wisps of gas from a supernova that took place in the constellation Cygnus (Swan) form the Veil Nebula.

Supernova 1987A

Supernovas are so bright that they can be spotted not only in our own galaxy but in other galaxies as well. In February 1987, a supernova erupted in the nearest galaxy to our own, the Large Magellanic Cloud. It was so bright that it could easily be seen with the naked eye. Astronomers reckoned that it was about 250 million times brighter than the Sun.

Supernova 1987A was the brightest exploding star seen in our skies since 1604 — before the telescope was invented. This time astronomers around the world trained their telescopes on the star to follow the progress of the explosion. A few years later the Hubble Space Telescope began observing what remained of the

Crushed

All that remains of a massive star after it has exploded as a supernova is the collapsing core. What happens next depends on the core's mass.

If its mass is up to about three times the mass of the Sun, the core shrinks to form a tiny body called a neutron star. Here is what happens.

As the core collapses under its own weight, it gets smaller and smaller. The smaller it gets, the stronger its gravity becomes, and the smaller and denser it becomes. In only seconds, it shrinks to a body only about 15 miles (25 km) across. Its atoms have been crushed, and it is made up mainly of tiny particles called neutrons. We now call it a neutron star.

Superdense

A neutron star is an amazing body — it has the mass of the Sun squeezed into a ball the size of a city. It is therefore incredibly dense — just a pinhead of its matter would weigh a million tons.

A neutron star also spins around rapidly. Its powerful magnetic field spins with it. Particles in the space around the star get caught up in the spinning magnetic field and whirled around. They eventually escape into space as beams of radiation.

The heavenly lighthouse

Because a neutron star spins, its beams sweep around in space like the beams from a lighthouse. From Earth, we observe them as flashes, or pulses, of radiation as they sweep past us. These pulses are as regular as the ticking of a clock.

beam of
radiation

neutron
star

Left: As a neutron star spins around, its beams sweep around, too.

Below: The beams from a rotating neutron star sweep through space (1). We see a pulse as it sweeps past Earth (2), before continuing on its way (3).

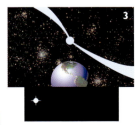

65

Little Green Men

Radio astronomer Jocelyn Bell at Cambridge, England, discovered the first pulsing signals from space in 1967. At first no one knew what caused them. It was even suggested that they were signals sent by intelligent beings, so they wre dubbed LGMs, standing for Little Green Men.

Soon many more pulsating sources, or pulsars, were detected, and it was realized that they were rapidly spinning neutron stars.

Hundreds of pulsars have now been discovered. Many spin around relatively slowly, about once a second. But some spin around more than 600 times a second. Just think of it — a body the size of a large city spinning around hundreds of times in a second.

Into a hole

A very big star doesn't end up as a neutron star. If the mass of its core is more than about three times the mass of the Sun, it will become an object called a black hole.

In the very heavy core of the big star, gravity is incredibly strong. It pulls all the matter in the core together with such force that it gets crushed.

Astronomers believe that the core eventually gets crushed into a tiny point. All that remains is a region around it that has a fantastically powerful gravitational pull. Anything that strays into this region is swallowed

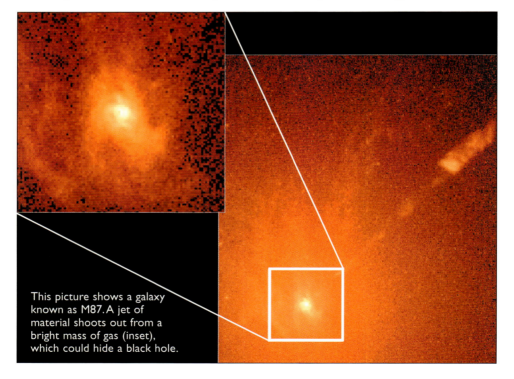

This picture shows a galaxy known as M87. A jet of material shoots out from a bright mass of gas (inset), which could hide a black hole.

up — even light rays. It is because it swallows light that such a region is called a black hole.

Discovering black holes

Because light can't escape from a black hole, we can't see it. But we can detect it in other ways. Like the star it came from, the black hole spins around. Nearby matter attracted by the black hole's gravity is dragged around too before being sucked inside. It forms a swirling disk of matter.

As it swirls furiously around, this matter gets incredibly hot, reaching temperatures of 200 million degrees F (100 million degrees C) or more. This makes it give off energy as X-rays, which can be detected from Earth.

Astronomers have picked up a powerful source of X-rays in the constellation Cygnus, called Cygnus X-1. They think that it comes from a swirling disk around a black hole.

The Cygnus X-1 black hole is part of a binary, or two-star, system. The other star is a blue supergiant. Astronomers think that the black hole is gradually pulling matter from this star into it. They reckon that this black hole is about 35 miles (60 km) across and has a mass about 10 times the mass of the Sun.

When there is a black hole in a binary system, it gradually draws off matter from the other star. The matter forms a fast spinning disk before being swallowed up.

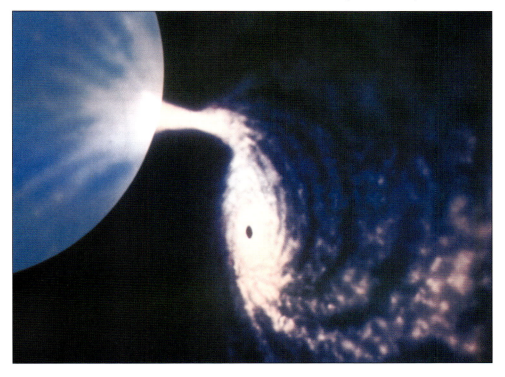

GALAXIES OF STARS

The stars are not scattered evenly throughout space. They are found clustered together in great star "islands", with empty space in between. These star "islands", or galaxies, contain billions of stars and measure thousands of light-years across.

In powerful telescopes, astronomers can see galaxies in almost every direction they look in space. In some parts of the sky they can see hundreds, even thousands of galaxies gathered together in clusters. Billions of galaxies make up the universe.

Most galaxies lie very far away. In small telescopes, they look like fuzzy blobs, rather like the gas and dust clouds we call nebulas. And astronomers once thought that they were nebulas. As telescopes improved, many of these "nebulas" were seen to have a spiral shape. Some astronomers thought that these spirals were located in our own galaxy, while others thought that they might be located outside it.

In 1919, a US astronomer named Edwin Hubble began investigating the spiral nebulas at the Mount Wilson Observatory outside Los Angeles. He used what was then the world's most powerful telescope, the Hooker. It had a light-gathering mirror 100 inches (2.5 metres) in diameter.

In this telescope, Hubble saw for the first time individual stars in the large spiral nebula in Andromeda. Among them he found variable stars called Cepheids. From the way the brightness of Cepheids changes, astronomers can figure out how far away they are.

Hubble worked out that the Cepheids he spotted in the nebula in Andromeda were nearly 1 million light-years away. This put them far beyond our own galaxy. The nebula had to be a separate star system — a separate galaxy. Since then, astronomers have found that the Andromeda Galaxy is more than twice as far away as Hubble calculated, at a distance of some 2.3 million light-years.

Left: Edwin Hubble

Opposite: Spiral galaxies like this are found throughout space. They can contain hundreds of billions of stars and measure hundreds of thousands of light-years across.

Classifying the Galaxies

Edwin Hubble found that galaxies came in many different sizes but usually had similar basic shapes. So he decided to classify galaxies by their shape, and we still follow his method today. There are two main kinds of galaxies — spirals and ellipticals. The others have no definite shape and are known as irregulars.

Maybe as many as three quarters of all galaxies are spirals. They are shaped rather like pinwheels. They have a central bulge of stars, surrounded by a disk of stars, gas, and dust. Within the disk, most of the stars are young and lie mainly on curved "arms" that give the galaxy its spiral shape.

In some spiral galaxies, the arms are close together, while in others they are quite open. Some galaxies have a bar of stars through the central bulge and are known as barred-spiral galaxies.

In a spin

Spiral galaxies spin around bodily in space, like a whirling pinwheel. You can tell in which direction they spin by looking at the curved arms. They trail behind the direction of spin like the streams of water from a garden sprinkler.

The stars in the disk circle within the disk region. It takes them millions of years to complete a full circle. But the stars within the bulge at the centre of a galaxy circle in many directions. Great globe-shaped clusters of older stars circle outside the bulge. These globular clusters travel within a hugesphere around the whole galaxy, called the halo.

Galaxies can have many shapes. Spherical (**1**) and egg-shaped (**2**) ones are classed as ellipticals. Others can be ordinary spirals (**3**) or barred spirals (**4**).

70

Elliptical galaxies

Elliptical galaxies can be round or oval in shape. They don't have the curving arms of spirals. Also they contain no dust and have no new stars forming in them. They are made up mainly of older stars.

The biggest galaxies in the universe are ellipticals. They are found in the centres of the great clusters of galaxies that exist in space. They probably got so large by gobbling up smaller galaxies that strayed too close to them.

When galaxies collide

Astronomers have found plenty of evidence that galaxies periodically bump into one another. When galaxies collide, their stars don't generally smash into one another. They pass each other by like ships in the night. This is because galaxies are mostly empty space, with the stars scattered about with plenty of space between them.

The Hubble Space Telescope provided this head-on view of galaxy NGC 4013 (above) and two galaxies colliding (below).

The Milky Way

On a clear moonless night, you can often see a faint, misty band of light arching across the sky. We call it the Milky Way. When you look at it in binoculars or a telescope, you see that it is made up of countless stars, seemingly packed close together.

What is the Milky Way? It is a view of our own galaxy from the inside. That is why we also call our galaxy the Milky Way, or sometimes just the Galaxy.

Where the Sun is

The Sun and all the other stars we see in the sky belong to the Milky Way Galaxy. In all, the Milky Way contains something like 100 billion stars. It seems to be a typical spiral galaxy, with a central bulge and a surrounding disk. It seems to have a faint bar through the centre, so maybe it is a barred-spiral galaxy.

The Milky Way Galaxy measures about 100,000 light-years across. The Sun sits on one of the curved arms that come out of the bulge, about 25,000 light-years from the centre.

In a halo

As in other spirals, globular clusters circle outside the bulge in a spherical halo that surrounds the whole galaxy. Nearly 150 of these clusters are known, the largest containing several million old stars.

The Milky Way rotates slowly in space. The stars take different periods of time to circle around the centre, depending on how far out they are. The Sun, for

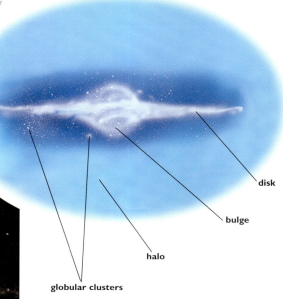

disk

bulge

halo

globular clusters

Above: Our galaxy has a bulge of stars in the centre and other stars on curved arms in the disk. All around it is a great sphere, or halo, of faint matter.

Left: This picture of our galaxy was produced from images sent back by a satellite.

72

example, takes about 225 million years to circle once around the galaxy.

Galactic Neighbours

Most galaxies lie so far away that we can see them only in telescopes. But there are three that we can see with the naked eye.

In the far southern hemisphere, you can spot two faint misty patches. The larger is called the Large Magellanic Cloud (LMC), the smaller, the Small Magellanic Cloud (SMC). They are named after the Portuguese navigator Ferdinand Magellan. He led the first expedition to sail around the world, setting out in 1519. He sailed across the southern oceans, from where the two Clouds could be seen.

Telescopes show that the LMC and SMC Clouds are really separate galaxies. They are in fact the nearest galaxies to the Milky Way. The LMC is closest, at a distance of only 170,000 light-years. It is a small galaxy, with only about one-third of the diameter of our own galaxy. The SMC is only about half the size of the LMC.

The great spiral

The other galaxy that we can see with the naked eye lies in the northern hemisphere in the constellation Andromeda. It was once called the Great Spiral in Andromeda, and great it is. It is one-and-a-half times bigger across than our own galaxy and it contains billions more stars.

The only reason we can see the Andromeda Galaxy in the night sky is because it is so large, for it lies an incredible 2.3 million light-years away. It the farthest object that we can see with the naked eye.

Above: This Hubble Space Telescope picture shows the Tarantula Nebula, a spidery-looking gas cloud in our neighbouring galaxy, the Large Magellanic Cloud.

Left: The Andromeda Galaxy is a spiral like our own galaxy, but it is much bigger. Close by are two smaller elliptical galaxies, which are linked to it by gravity.

Left: The Large Magellanic Cloud is one of the satellites of our own galaxy.

Below: A cluster of distant galaxies. Clusters like this can be seen in many regions of space — some contain thousands of galaxies.

Clustering Together

The Magellanic Clouds, the Andromeda Galaxy, and the Milky Way are neighbours in space. They form part of a cluster of galaxies in our part of the universe known as the Local Group. All the galaxies in the Group are bound together by gravity.

The Magellanic Clouds circle around our galaxy like space satellites around Earth, and they are known as satellite galaxies. They are gradually coming closer, and one day our galaxy will swallow them up. The Andromeda galaxy has smaller satellite galaxies too.

Scattered about

In all there are over 30 galaxies in the Local Group, and they are scattered over a region of space about 7 million light-years across. The Andromeda Galaxy is the biggest, followed by the Milky Way; then comes M33, in the constellation Triangulum (Triangle). These three galaxies are the only spirals. The rest are small elliptical or irregular galaxies.

The more distant galaxies gather together in clusters too. Some of these clusters contain thousands of individual galaxies. An example is the cluster in the constellation Virgo, which contains at least 1,000 galaxies.

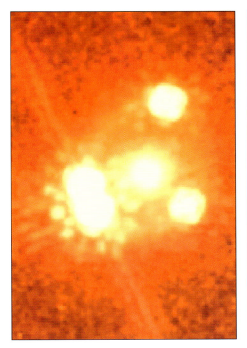

Left: A distant quasar, which pours out fantastic energy into space as light and other kinds of radiation.

Active Galaxies

Most galaxies give off much of their energy as light. The amount of energy they give off is what would be expected for a group of billions of stars. But a few galaxies give off much more energy than expected. We call them active galaxies.

Some active galaxies pump out their exceptional energy as light. Others pour it out most of their energy in the form of radio waves, or X-rays.

Amazing quasars

Astronomers also pick up powerful radio beams from incredibly distant objects called quasars. These objects lie billions of light-years away, yet are still visible as faint stars. They seem to produce as much energy as a thousand normal galaxies.

Where do active galaxies get their energy? Certainly, the ordinary nuclear processes in stars couldn't provide it. So astronomers think that the energy comes from black holes in the heart of the galaxies. A black hole produces energy from the ring of matter swirling round it.

The black holes in active galaxies would have to have millions of times the mass of our Sun to produce the amazing energy given out.

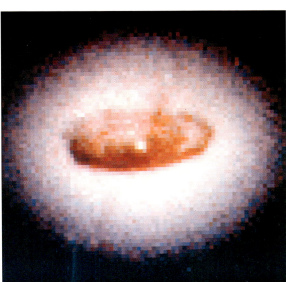

Left: This donut-shaped ring of dusty matter is the centre of a giant active galaxy. Astronomers think that a black hole lurks inside it.

THE MIGHTY UNIVERSE

All the planets and stars, the galaxies and clusters, and space itself make up the universe. Astronomers think they know what the universe is like and how it began in a huge explosion long ago. But they're not sure what will happen to it in the future.

All things that exist — rocks, air, living things, and the heavenly bodies floating in space — make up the universe. Another name for the universe is the cosmos.

The universe as we know it today is made up mainly of bodies floating in space — planets and their moons, stars, and galaxies. These bodies are widely scattered, and most of the universe is just empty space.

In early times, people had no idea what the universe was like. They thought that our world was the centre of the universe, and had different ideas about what it was like. In ancient India, for example, people thought that the world was carried on the back of four elephants. In turn, the elephants stood on the shell of a huge tortoise. And the tortoise itself was carried by a snake swimming in a vast ocean.

The ancient Greeks knew Earth was round. They believed it lay in the centre of a great celestial, or heavenly, sphere, which spun around. The stars were stuck to the inside of the sphere. The Sun, the Moon and the planets circled Earth closer in. The celestial sphere was the universe.

In 1543 an astronomer named Copernicus turned ideas on their head by suggesting that Earth circled around the Sun, and not the other way round. The other planets circled round the Sun as well, forming a family of bodies called the solar system. The universe then seemed to be the solar system.

Over time, astronomers realized that the Sun was merely a star like the other stars in the night sky. Then the universe became the great star system we call our galaxy.

This takes us to the beginning of last century. Then astronomers found that there were many more galaxies around. So the universe seemed to be made up of galaxies and clusters of galaxies floating in space. And that is more or less what astronomers believe today.

Opposite: The universe is bigger than we can ever imagine. It is mostly empty space, with galaxies of stars, dust, and gas scattered here and there.

1

Astronomers think that a violent event they call the Big Bang (**1**) created the universe and began its expansion. For a while it was incredibly hot (**2**).

2

The Expanding Universe

We saw earlier that astronomers can tell how a star is moving by studying the light it gives out. They look at the way the dark lines in its spectrum shift towards the blue end or the red end. A blueshift shows that the star is travelling toward us, a redshift that the star is travelling away.

Astronomers can tell whether a galaxy is travelling toward or away from us in a similar way. When they study the light from galaxies, they find that it almost always shows a redshift. In other words, nearly all the galaxies are rushing away from us. And the farther they are away, the faster they seem to be travelling.

It seems as if the whole universe is getting bigger, or expanding. If this is true, the universe must have been smaller in the past. Working backward, there must have been a time when everything in the universe was packed together in one place. Then something must have happened that set the universe expanding.

Most astronomers agree that something like this actually happened.

The universe came into being as a point, and a great explosion began its expansion. They call this explosion the Big Bang. It probably happened about 15 billion years ago, but astronomers don't know exactly when.

After the bang

The astronomers who study the universe as a whole and how it has developed are known as cosmologists. Incredibly, they think that they know what has happened to the universe nearly from the instant it was born.

To start with, the universe was tiny and incredibly hot — with temperatures of trillions of trillions of degrees. It consisted only of energy, because no matter could exist at such fantastic temperatures. But as the universe expanded rapidly, it also cooled down rapidly.

78

Above: We find galaxies everywhere when we look deep into space. The farthest ones lie more than 10 billion light-years away.

Below: Then, as the universe got bigger, it cooled down, and matter started to form (**3**). Some 15 billion years after the Big Bang, we find the universe as it is today (**4**), still expanding.

As it cooled down, tiny bits of matter — particles — began to form. Over time, these particles joined together to form simple atoms of hydrogen. And later the atoms formed into clouds. It was from these clouds that the first stars were born. In time, the stars began to collect together into galaxies. No one is certain when this took place. But it may have happened before the universe was 1 billion years old.

The Universe Today

Fifteen billion years after the Big Bang, we come to the universe as we find it today. It has expanded to become bigger than we can ever imagine. The farthest objects we can see in our most powerful telescopes seem to lie nearly 15 billion light-years away. (In miles this is 10 followed by 22 zeros, or in kilometres 16 followed by 22 zeros.)

Looking at things in another way, the light from these objects has been travelling toward us for nearly 15 billion years. So we are seeing them as they were shortly after the Big Bang, not as they are today.

3

4

Structure of the Universe

The diagram on the previous page reminds us of the structure of the universe as we find it today, from planets and stars, galaxies and clusters, to enormous superclusters. Curiously, the superclusters seem to form kinds of thin sheets around otherwise empty spaces, called voids.

The End

We think we know what the universe was like in the very distant past, and what it is like today. But what will happen to it in the very distant future?

There seem to be three main possibilities. One is that the universe will continue to expand forever until it eventually runs out of energy. Another idea is that the universe will eventually stop expanding, leaving it at a certain size.

The third idea is that the expansion will stop expanding and then go into reverse — it will start to shrink. It will carry on shrinking until it ends up, as it began, as a tiny point. Astronomers say there would be a Big Crunch.

Dark matter

The thing that will decide how the universe will end is gravity. If there is enough matter in the universe, there will be enough to stop the universe expanding and even make it shrink. If there is not enough matter, the universe will expand forever.

There is not enough matter in all the stars and galaxies we can see to stop the universe from expanding. But astronomers believe that there is a lot of matter in the universe that they can't see. It is called dark matter. They can't see it, but they can detect ts gravity.

Astronomers have estimated how much matter there could be in the universe. But it does not seem to be enough to hold back the fleeing galaxies. So it seems as if the universe could go on expanding forever.

Earth is a rocky planet that circles the Sun.

We live in cities on Earth's land areas, or continents.

The universe in scale: These images give an idea of how we humans fit into the universe. The bottom line is that in the universe as a whole, we are not very important at all! And the planet we live on is but a grain of salt in a vast ocean of space.

Above: One of the many clusters of galaxies found throughout the universe. It is a particularly interesting cluster, because it acts like a lens to bend light from a galaxy behind it. We see this galaxy as a series of blue arcs.

Billions of galaxies cluster together to make up the vast universe.

The Sun and billions of other stars belong to a galaxy, or great star island in space.

The Sun's family of planets and other bodies travel through space together.

81

Exploring the Night Sky

In this chapter we explore the fascinating world of astronomy — the scientific study of the heavens. Astronomy began with the earliest humans to look in the night sky at the stars and wonder about what they saw there. With the first great civilizations of the Middle East and elsewhere, astronomy became a subject of scientific and religious study.

Like our early ancestors, we begin our study of astronomy with simple stargazing, using only our eyes to observe what happens in the heavens. We will learn how to recognize the constellations, those glittering patterns of bright stars in which our ancestors saw shapes and meaning. In Section Three of the book, a set of star maps is included so that you can identify the constellations as they appear in the night sky in the different seasons.

Stargazing with only our eyes gives just a hint of what the universe of stars, planets, and galaxies is like. Looking at the heavens through binoculars or a telescope brings more spectacular sights into view, delighting the eye and exciting the imagination — sights such as colourful clouds of glowing gas, and clusters of stars sparkling like jewels.

From lofty mountaintop observatories, today's astronomers peer at the distant heavens with giant telescopes, able to spy objects whose light has been

The Hubble Space Telescope captured this beautiful image of Galaxy NGC 4214. Located some 13 million light-years from Earth, NGC 4214 is currently forming clusters of new stars from its interstellar gas and dust.

travelling towards Earth from unimaginable distances for billions of years.

Even stranger sights are viewed by space telescopes such as the Hubble — newborn stars and planetary systems, dying suns blasting themselves apart. Nearer home, unmanned probes explore our own planetary system — spying volcanoes erupting on Jupiter's moon, Io, finding evidence of flowing water on Mars.

Though astronomy is one of the oldest sciences, there is nothing old-fashioned about the way astronomy is practiced today. Practically every day brings some new insight from today's astronomers about the way the universe began and how it works.

THE DAWN OF ASTRONOMY

The beginnings of astronomy can be traced back more than 5000 years, to the very beginnings of civilization. Until only about 400 years ago, astronomy had to rely only on observations that could be made with the eye alone. Then the invention of the telescope opened up a new window on the universe.

In all likelihood, prehistoric people looked at the night sky and marvelled at its beauty. Humans probably did not start studying the heavens carefully until about 10,000 years ago, when they began farming and living a more settled life. They may have used their observations of the night sky to determine such things as the change of seasons and when was the right time to sow crops. The oldest known astronomical records date to about 3000 BC, when writing developed in the Middle East.

There, the Babylonians recorded astronomical observations on clay tablets in cuneiform writing (a system of writing with wedge-shaped marks pressed into soft clay, or cut into stone). The Egyptians kept records in the form of hieroglyphics, or picture writing.

These early records show that, in the Middle East at least, astronomy was already well advanced.

The Egyptians worked out an accurate calendar, based on a year of 365 days, much like the one we use today. The ancient Chinese developed an accurate calendar, too. They also noted unusual events in the heavens — they recorded eclipses as far back as 4000 BC.

The prehistoric people in Europe must have had some knowledge of astronomy, too. They built huge stone circles, such as Stonehenge in England, which dates from about 2000 BC. Stonehenge seems to have been a kind of observatory. Its huge stones, or megaliths, seem to have been laid out with a specific purpose in mind. Certain stones lined up at certain times of the year to point to where, for example, the Sun rose on the summer solstice — the longest day of the year.

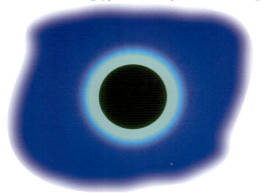

Left: Total eclipses of the Sun scared ancient peoples. But as early as 585 BC, the Greek astronomer Thales was able to predict when eclipses would occur.

The ancient Egyptians laid out the great pyramids with great precision, aligning them with prominent stars.

The ancient observatory Stonehenge as it is today. It is located near the town of Amesbury on Salisbury Plain, England. The most striking features are the trilithons, tall structures built of two massive sandstone blocks with another block on top.

Greek Astronomy

The civilization developed by the ancient Egyptians along the Nile River thrived for more than 2,000 years. Weakened by internal squabbles, it was invaded and occupied by a succession of foreign powers, beginning in about 945 BC. By then another powerful civilization was flourishing in the Mediterranean — ancient Greece.

The ancient Greeks built on the knowledge of the heavens inherited from the Babylonians and Egyptians. From 600 BC, Greek philosophers began trying to understand what happened in the heavens and how the universe worked. They laid the foundations for scientific astronomy.

The Egyptians had believed in a flat Earth, but Greek philosophers such as Aristotle (fourth century BC) thought otherwise. The Earth had to be round, they argued. As evidence, they pointed out that the shadow of the Earth on the Moon during an eclipse of the Moon was curved.

The Fixed Earth

Aristotle taught that the round Earth was fixed at the centre of the universe and that the Sun, Moon, and planets circled around it.

About a century after Aristotle, Aristarchus determined that the motions of the heavenly bodies could just as easily be explained if Earth was a planet and circled around the Sun. Aristarchus was right, of course, but

Right: The universe according to Ptolemy. Earth is in the centre, with the moon (Luna), Sun (Solis), and planets circling around it.

Below: Ptolemy introduced the idea of deferents and epicycles to explain the peculiar movement of the planets in the sky. Each planet, he said, moved in a circle (epicycle) around a point that moved in a circle (deferent) around Earth.

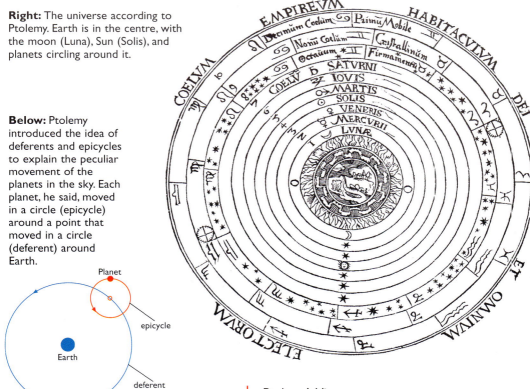

his ideas were too far ahead of their time.

The Great Observer

The greatest Greek observational astronomer was undoubtedly Hipparchus, who lived in the second century BC. He drew up a catalog charting the position and brightness of almost 1000 stars. He devised the magnitude system by which astronomers still classify star brightness: the brightest stars we can see are of the first magnitude, and the faintest stars are of the sixth magnitude; other stars have magnitudes in between.

Ptolemy's View

Hipparchus's work would have been lost were it not for the last great Greek astronomer, Ptolemy of Alexandria, who lived in the 2nd century AD. He included it in a book he wrote that summed up all the astronomical and scientific knowledge of the time. His book has survived in an Arab translation and is usually called the *Almagest*.

Ptolemy set out the accepted view of the universe — a spherical Earth, around which the heavenly bodies travelled in circles. The stars were fixed to the inside of a great celestial (heavenly) sphere surrounding Earth. This Earth-centred view of the universe is known as the Ptolemaic system.

In the East

Much of the knowledge obtained by the ancient Greeks was passed on to the Roman Empire, which at its peak stretched over much of Europe, North Africa, and the Middle East. In the 5th century, the empire fell into decline and came under attack from barbarian tribes.

With the fall of Rome, a period known as the Dark Ages began in Europe. Interest in learning and science — including astronomy — declined, and much earlier knowledge was lost and forgotten.

During this time, however, astronomers in other parts of the world were making advances. In the east, in India, astronomers were building primitive observatories. Arab astronomers were building new instruments, such as the astrolabe, to measure the positions of stars. By the 9th century,

Above: An Arabian astrolabe, used to measure the angles of stars above the horizon. An astronomer trained the sighting bar on a star, then read the angle above the horizon from a scale around the edge.

the city of Baghdad (in what is now Iraq) was the site of the finest observatory the world had ever known.

In the West

At the same time, astronomy was flourishing in Central America, a part of the world unknown to either European, Indian, or Arab astronomers. There, the Mayan civilization was at its peak. Mayan astronomers were meticulous observers who left detailed records in manuscripts and carved in stone. They kept particularly accurate records of the movements of Venus, by which they checked their calendar. They could also predict when eclipses would take place.

Nicolaus Copernicus

European Astronomy Reborn

By the 15th century, a general revival of learning was underway in Europe. This period became known as the Renaissance. People began developing new ideas and challenging age-old beliefs. Among them was the Polish astronomer Nicolaus Copernicus.

Copernicus questioned Ptolemy's Earth-centred view of the universe. He realized that what seemed to be odd movements of the planets in the sky

Left: An armillary sphere was an early astronomical device for representing "great circles" in the heavens, such as the meridian (north-south circle) and ecliptic — the path the Sun seems to take through the heavens during the year.

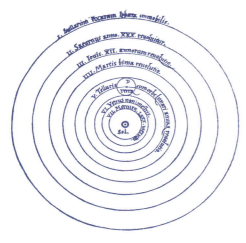

Above: A sketch by Copernicus showing his solar system, centreed on the Sun (Sol). Earth (Terra) becomes just another planet.

Below: Tycho Brahe's famous observatory Uraniborg ("Castle of the Heavens") on the island of Hven, Denmark, built in 1576. It boasted the finest instruments of the time.

could be better explained if Earth circled the Sun. Thus, the old idea of a sun-centred, or solar, system was reborn.

Copernicus did not publish his ideas until he lay on his deathbed in 1543. He knew that they would upset the accepted view of the universe held by the Catholic Church, which taught that Earth was the centre of the Universe.

Convincing Evidence

In the years that followed, the Church did punish people who agreed with Copernicus's ideas about the solar system. Eventually astronomers provided convincing evidence that it was right.

One was a German astronomer named Johannes Kepler. He used the very accurate observations made by the Danish astronomer Tycho Brahe to work out exactly how the planets moved through the sky. He proved that their motions could be explained precisely if they travelled around the Sun in oval, or elliptical orbits. He published this finding in 1609 as the first of his celebrated laws of planetary motion.

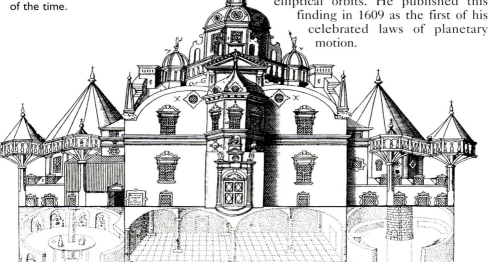

THE CONSTELLATIONS

The stars shine down on us from the great dark dome of the heavens. The brightest ones glow like beacons and guide us around the night sky. Long ago, humans thought they detected patterns in the stars. They gave these shapes and patterns the names by which they are still known today — the constellations.

The ancient astronomers saw virtually the same constellations as we see today. Even over thousands of years, the stars of the constellations barely change position.

People began naming the constellations more than 5,000 years ago. They named them after the figures, shapes, and patterns they thought they saw — a lion here, a swan there.

Astronomers often use Latin names for the constellations, so lion becomes the constellation Leo, and the swan becomes Cygnus.

There are interesting stories about most of the constellations. The Greeks named nearly 50 of them, often relating them to their myths, or traditional stories. Birds, fish, sea monsters, even everyday objects found their way into the sky, each with a tale to tell.

Some of the brightest stars in the constellations have their own names. For example, the brightest star in the whole heavens is in the constellation Canis Major (Great Dog). It is named Sirius and is more popularly called the Dog Star.

The third-brightest star in the heavens is in the constellation Centaurus (Centaur). It is usually called Alpha Centauri. Astronomers use this kind of name all the time, using a Greek letter followed by the constellation name. Alpha is the first letter in the Greek alphabet, which tells us that Alpha Centauri is the brightest star in Centaurus. Beta Centauri is the second-brightest star, and so on.

Left: The constellation Taurus is depicted as the head of a charging bull. The bright reddish star Aldebaran marks the bull's eye.

Right: Globe-shaped masses of stars like this are found in many of the constellations. Called globular clusters, they can contain hundreds of thousands of stars packed closely together.

In the Same Direction

The stars in the constellations look as if they are grouped quite close together in space. But they are actually immense distances from one another. The stars in a constellation only look close together because they happen to lie in the same direction in space when we view them from Earth.

For example, the nearest star to us in the constellation Orion is about 300 light years away from Earth. But the bright star Betelgeuse is over 600 light years away, and Rigel is over 800 light years away.

All Change

The constellations seem as if they never change. Their stars always seem to stay in the same positions, which is why we see the same constellations as the ancient Babylonians did thousands of years ago.

Astronomers tell us that the constellations are changing all the time —

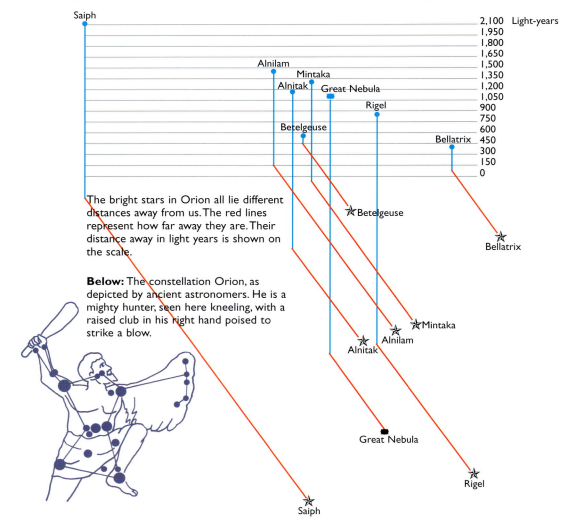

The bright stars in Orion all lie different distances away from us. The red lines represent how far away they are. Their distance away in light years is shown on the scale.

Below: The constellation Orion, as depicted by ancient astronomers. He is a mighty hunter, seen here kneeling, with a raised club in his right hand poised to strike a blow.

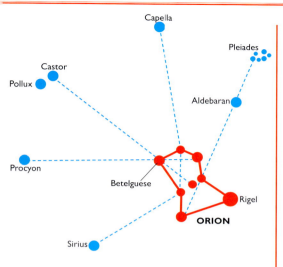

Capella
Pleiades
Castor
Pollux
Aldebaran
Procyon
Betelgeuse
Rigel
ORION
Sirius

Above: Orion is an excellent signpost to other stars, including the brightest star, Sirius.

but so little and so slowly that our eyes cannot detect the change. By studying light from the stars, astronomers know that they are all travelling very fast indeed. The reason we cannot see them move is because they are so very far away — trillions upon trillions of miles.

But if we were able to return to Earth millions of years from now, it would appear to us that the stars have moved. The constellations would no longer appear as the same patterns because the stars are all moving in different directions.

Guiding the Way

The constellations can help us find our way around the heavens, serving as signposts to other constellations, individual stars, and other features.

In the Northern Hemisphere, the Big Dipper is a good signpost. A line through the two stars at the end of the

Right: A dark celestial sphere appears to surround Earth. The stars seem to be stuck on the inside of the sphere.

dipper points to the Pole Star, Polaris. Polaris is also called the North Star because it is located in the heavens directly above the North Pole. When you find it you know you are looking north. Sailors have used the Pole Star to navigate at sea for centuries.

Stargazers in both the Northern and Southern Hemisphere can use another unmistakable constellation Orion as a guide. As the illustration shows, it can be used to locate a host of stars in other constellations.

The Celestial Sphere

Ancient astronomers believed that the stars were stuck on the inside of a great round ball that surrounded Earth. They called this ball the celestial, or heavenly, sphere.

We still use the idea of a celestial sphere even though we know that it does not actually exist. We imagine the sphere to be divided into two halves. One half covers the Northern Hemisphere of the world, and the other half the Southern Hemisphere. They meet along an imaginary line, which we call the celestial equator. It lies directly over the Earth's Equator.

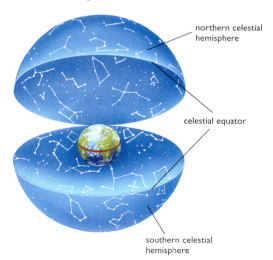

northern celestial hemisphere

celestial equator

southern celestial hemisphere

Northern Constellations
Constellations of the northern celestial hemisphere.

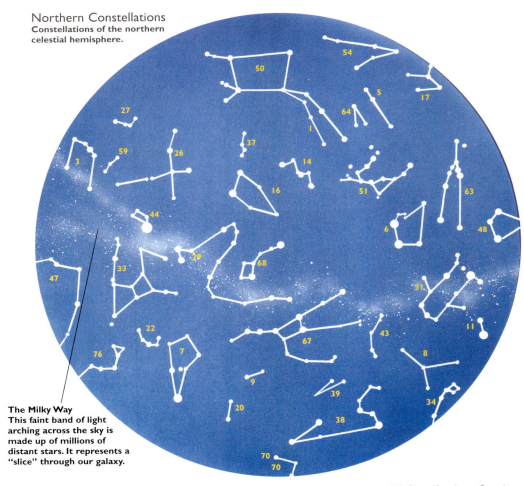

The Milky Way
This faint band of light arching across the sky is made up of millions of distant stars. It represents a "slice" through our galaxy.

1 Andromeda
2 Aquarius (Water-Bearer)
3 Aquila (Eagle)
4 Ara (Altar)
5 Aries (Ram)
6 Auriga (Charioteer)
7 Boötes (Herdsman)
8 Cancer (Crab)
9 Canes Venatici (Hunting Dogs)
10 Canis Major (Great Dog)
11 Canis Minor (Little Dog)
12 Capricornus (Sea Goat)

13 Carina (Keel)
14 Cassiopeia
15 Centaurus (Centaur)
16 Cepheus
17 Cetus (Whale)
18 Chamaeleon (Chameleon)
19 Columba (Dove)
20 Coma Berenices (Berenice's Hair)
21 Corona Australis (Southern Crown)
22 Corona Borealis (Northern Crown)
23 Corvus (Crow)
24 Crater (Cup)

25 Crux (Southern Cross)
26 Cygnus (Swan)
27 Delphinus (Dolphin)
28 Dorado (Swordfish)
29 Draco (Dragon)
30 Eridanus
31 Gemini (Twins)
32 Grus (Crane)
33 Hercules
34 Hydra (Water Snake)
35 Hydrus (Little Snake)
36 Indus (Indian)

Southern Constellations

Constellations of the southern celestial hemisphere.

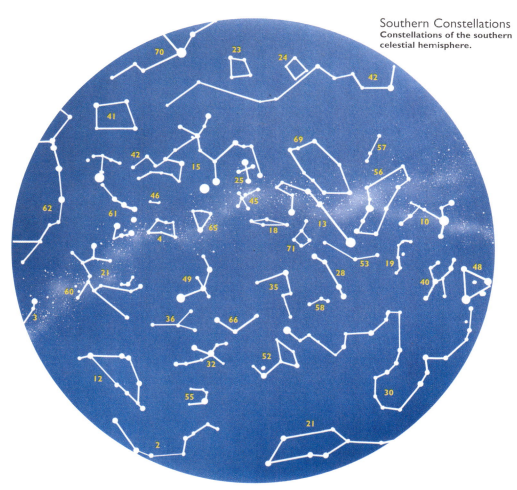

37 Lacerta (Lizard)
38 Leo (Lion)
39 Leo Minor (Little Lion)
40 Lepus (Hare)
41 Libra (Scales)
42 Lupus (Wolf)
43 Lynx (Lynx)
44 Lyra (Lyre)
45 Musca (Fly)
46 Norma (Rule)
47 Ophiuchus (Serpent-Bearer)
48 Orion

49 Pavo (Peacock)
50 Pegasus (Flying Horse)
51 Perseus
52 Phoenix (Phoenix)
53 Pictor (Painter)
54 Pisces (Fishes)
55 Piscis Austrinus (Southern Fish)
56 Puppis (Poop)
57 Pyxis (Compass)
58 Reticulum (Net)
59 Sagitta (Arrow)
60 Sagittarius (Archer)

61 Scorpius (Scorpion)
62 Serpens (Serpent)
63 Taurus (Bull)
64 Triangulum (Triangle)
65 Triangulum Australe (Southern Triangle)
66 Tucana (Toucan)
67 Ursa Major (Great Bear)
68 Ursa Minor (Little Bear)
69 Vela (Sails)
70 Virgo (Virgin)
71 Volans (Flying Fish)

The Constellations You See

Even if you spend every night of the year stargazing, you will never be able to see all 88 constellations. Because Earth is round, part of the sky is always hidden from view.

For example, if you live in Vermont, the Little Dipper and the Pole Star, Polaris, will be old friends. But you will never see Crux, the famous Southern Cross, which lies "down under" in the far south of the heavens. On the other hand, stargazers in South Australia will be familiar with Crux, but will never see the Little Dipper.

Looking at Latitude

In general, over the course of a year an observer in the Northern Hemisphere will be able to see all the constellations of the northern celestial hemisphere and some of those of the southern. Similarly, an observer in the Southern Hemisphere will be able to see all the constellations of the southern celestial hemisphere and some of those of the northern.

Exactly which constellations stargazers will see depends on their latitude, or how far away they are away from Earth's Equator.

Below: Because Earth is round, observers in different parts of the world see different views of the heavens.

Below: Going to a planetarium is a good way of learning about the stars and other heavenly bodies. In a planetarium, a special projector throws images of the night sky on a domed ceiling.

The Whirling Heavens

When you stargaze for any length of time, you will notice that the constellations seem to slowly change positions, moving across the sky from east to west. New stars seem to continually rise above the horizon in the east, while others set beneath the horizon in the west. The entire celestial sphere seems to spin around, which is what the ancient astronomers believed. But they were wrong.

It is Earth that spins around in space from west to east, and this makes the heavens seem to spin around us in the opposite direction.

Star Maps and Planisphere

Because Earth moves, which con-stellations you see and where they are in the sky depends on the time of the night at which you are viewing.

Right: A planisphere has a movable disc on top, which rotates over a map of the heavens. When you line up the observing time (on the movable disc) with the date (on the base), a view of the night sky at that time and date appears in the transparent window.

A very useful tool that all stargazers should take with them is a planisphere. This device shows the constellations visible in the sky at any time of the night on every night of the year. Because our view of the heavens depends on how far north or south of the Equator we are, different planispheres are made for different latitudes.

Planisphere shows the principal stars visible for every hour in the year for Latitude 42°N USA • Southern Europe • Northern Japan

Left: Circular star trails. You get a picture like this when you point your camera at the Pole Star and leave the camera shutter open for an hour or more. The stars make circular trails because Earth is spinning beneath them.

THE CHANGING HEAVENS

The night sky does not change only during the course of the night. It also changes gradually during the year. Month by month, different constellations sweep into view while others disappear, making the heavens in each season distinctly different.

Look south in the sky in the evening around the turn of the New Year and you will see the brilliant constellation Orion. Look in the same direction at the same time of night three months later and you will see the unmistakable shape of Leo (Lion). Orion will have all but disappeared over the western horizon. As the months go by, Leo moves on and is replaced by other constellations.

The constellations seem to come and go like this because Earth circles in orbit around the Sun during the year. Every month, Earth travels a little further in its orbit. This means that each night we look out onto a slightly different part of the heavens — of the so-called celestial sphere. After several months, our view is of quite a different part of the heavens. Therefore we see different constellations.

Because Earth always completes its orbit of the Sun in one year, each year we can see the same constellations in the same place in the sky at the same time of the year. In the next pages, we look at the constellations that appear in the heavens season by season — in spring, summer, fall, and winter. The main maps show the stars visible at about 10 p.m. to stargazers in the United States and other countries in the Northern Hemisphere.

The sky views shown are not exactly what all stargazers will see. What you will see depends on your latitude, or how far you are away from the equator. If you live in the Great Lakes region, for example, you will be able to see a little farther north, while someone who lives in Florida will be able to see a little farther south.

Stargazers in the Southern Hemisphere of the world have different views of the night sky. The constellations familiar to northern stargazers appear upside-down to them. In addition, the seasons are different as well. Spring in the northern hemisphere is fall in the southern hemisphere, winter in the north is summer in the south, and so on.

Opposite: The dazzling Pleiades star cluster, also called the Seven Sisters. It is one of the highlights of the constellation Taurus.

Spring Stars

By mid-April, the Big Dipper has reached its highest point in the heavens and is almost directly above the Pole Star, Polaris. In contrast, Cassiopeia is almost at its lowest point and sits on the northern horizon.

Two bright stars have risen above the horizon in the north-east. They are Deneb in Cygnus (Swan) and Vega in Lyra (Lyre). Deneb is much farther away than Vega but shines brightly because it has the brilliance of 50,000 Suns.

Cygnus lies in the Milky Way, which

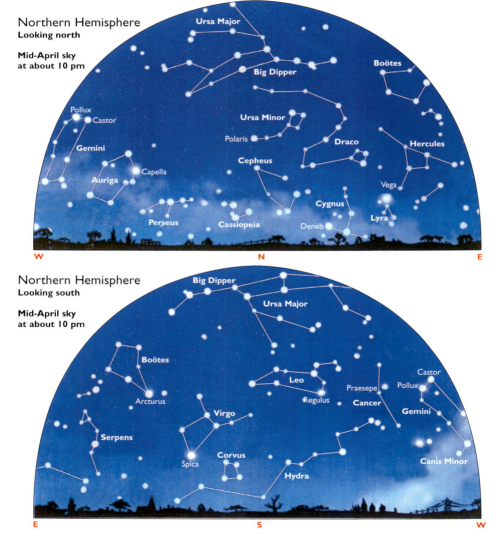

Northern Hemisphere
Looking north

Mid-April sky
at about 10 pm

Ursa Major
Big Dipper
Boötes
Pollux
Castor
Gemini
Ursa Minor
Polaris
Draco
Hercules
Capella
Cepheus
Auriga
Vega
Cygnus
Lyra
Perseus
Cassiopeia
Deneb

W   N   E

Northern Hemisphere
Looking south

Mid-April sky
at about 10 pm

Big Dipper
Ursa Major
Boötes
Leo
Castor
Praesepe
Pollux
Arcturus
Regulus
Cancer
Virgo
Gemini
Serpens
Corvus
Canis Minor
Spica
Hydra

E   S   W

at this time is located close to the horizon. Viewed through binoculars, the Milky Way is a delight — sprinkled with millions of close-packed stars, clusters, and shining gas clouds.

Looking South

The easiest constellation to recognize is Leo (Lion). Its brightest star, Regulus, and the curve of stars above it form a pattern aptly named the Sickle. To the west are the two brightest stars of Gemini (Twins), Castor and Pollux.

In between Leo and Gemini, there is a little group of stars in Cancer (Crab) that you can see with the naked eye. This cluster is called the Beehive because the scattered stars look like bees buzzing around a hive.

Towards the east, two bright stars shine out of a generally dull part of the sky. One is Arcturus in Boötes (Herdsman), which is the brightest star in the northern hemisphere and the fourth-brightest overall. The other brilliant star is Spica, the leading star in Virgo (Virgin).

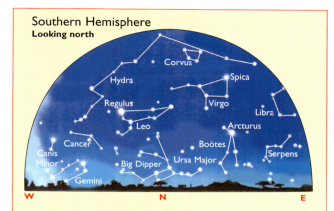

Southern Hemisphere
Looking north

When it is spring in the Northern Hemisphere of the world, it is fall in the Southern Hemisphere. Looking north in the April sky, southern observers see most of the constellations that northern observers see when they look south, but upside-down.

Leo is prominent in the mid-sky. Its brightest star, Regulus, forms a conspicuous triangle with two other bright stars — Arcturus and Spica.

This month southern observers have a rare chance to see the Big Dipper, which appears low down on the northern horizon.

Looking South

The skies are, as ever, dazzling. Crux (Southern Cross) is in the mid-sky almost due south. It is one of the many gems in the bright arc of the Milky Way. The two brightest stars in the whole heavens, Sirius and Canopus, shine like beacons in the south-west.

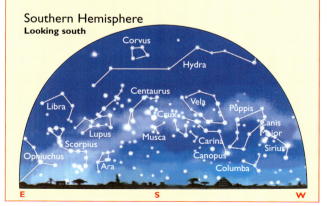

Southern Hemisphere
Looking south

101

Summer Stars

If you look north at 10 p.m. on an evening in mid-July, the Big Dipper is descending, while Cassiopeia is climbing. Pegasus and Andromeda are rising in the east, bringing into view the misty patch known as the Great Nebula in Andromeda. It is actually a galaxy, the light of which has been travelling toward us for more than 2,000,000 years. It is the most distant heavenly body we can see with our eyes alone. A powerful telescope is needed to see its individual stars.

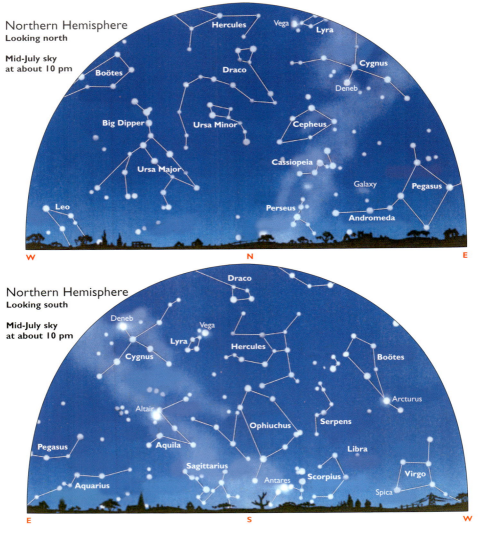

Northern Hemisphere
Looking north

Mid-July sky
at about 10 pm

Hercules · Vega · Lyra · Cygnus · Boötes · Draco · Deneb · Big Dipper · Ursa Minor · Cepheus · Ursa Major · Cassiopeia · Galaxy · Pegasus · Leo · Perseus · Andromeda

W · N · E

Northern Hemisphere
Looking south

Mid-July sky
at about 10 pm

Draco · Deneb · Vega · Lyra · Hercules · Boötes · Cygnus · Arcturus · Altair · Serpens · Ophiuchus · Pegasus · Aquila · Libra · Sagittarius · Virgo · Antares · Scorpius · Aquarius · Spica

E · S · W

Looking South

A trio of bright stars hits the eye. They are Altair in Aquila (Eagle), Deneb in Cygnus (Swan), and Vega in Lyra (Lyre). These beacon stars are a distinguishing feature of summer skies and form what is called the Summer Triangle.

Low on the southern horizon, this month observers can glimpse two of the most spectacular constellations of the southern hemisphere. They are Sagittarius (Archer) and Scorpius (Scorpion). Both straddle the Milky Way and are rich in star clouds and clusters. The Milky Way appears dense here; when we look at Sagittarius we are looking into the heart of the galaxy.

In Scorpius, observers should be able to spot the red star Antares, which marks the Scorpion's heart. Stargazers in the southern states may see the curve of bright stars that trace out the animal's deadly stinging tail.

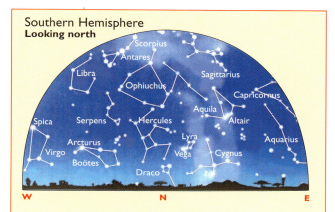

Southern Hemisphere
Looking north

In the Southern Hemisphere, July is a winter month. As the stargazer looks north, three bright stars appear low in the sky. These are the three stars that form the celebrated Summer Triangle in the Northern Hemisphere. They are Deneb, Altair, and Vega. Deneb marks the tail end of the Swan, which here is flying nearly vertically upward through the sky along the Milky Way.

Toward the west appear two constellations very familiar to northern stargazers — Hercules and the kite-shaped Boötes (Herdsman) with its brilliant star Arcturus.

Looking South

The brilliant Milky Way rises almost vertically in the sky. Crux (Southern Cross) is prominent near the horizon, surrounded by the brilliant stars of Centaurus (Centaur). This constellation's brightest star, Alpha Centauri, is one of the nearest stars to us, being only a little over 4 light years away.

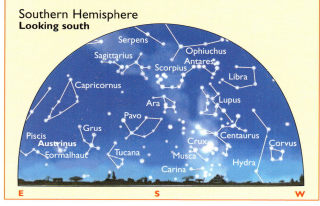

Southern Hemisphere
Looking south

Fall Stars

With the coming of fall, the evening skies darken noticably. The Milky Way is becoming more prominent. In mid-October, it spans the sky from east to west.

As the stargazer looks north, the Big Dipper is reaching its lowest point in its perpetual orbit around the Pole Star, Polaris, and now sits on the horizon. Cassiopeia, by contrast, is now riding high near the top of the arc of the Milky Way.

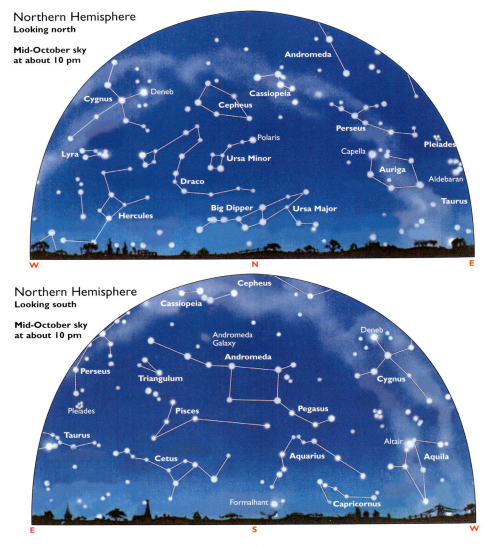

Northern Hemisphere
Looking north

Mid-October sky
at about 10 pm

Andromeda

Cygnus · Deneb · Cassiopeia
Cepheus

Perseus

Pleiades

Lyra · Polaris · Capella
Ursa Minor · Auriga

Draco · Aldebaran

Hercules · Big Dipper · Ursa Major · Taurus

W · N · E

Northern Hemisphere
Looking south

Mid-October sky
at about 10 pm

Cepheus
Cassiopeia

Andromeda
Galaxy

Deneb

Andromeda

Perseus · Triangulum · Cygnus

Pleiades · Pisces · Pegasus

Taurus · Altair

Cetus · Aquarius · Aquila

Formalhant · Capricornus

E · S · W

Toward the west, Cygnus (Swan) appears in the mid-sky, where its swan shape — long thrusting neck and outstretched wings — can be best appreciated. In the east, Taurus (Bull) has risen, its lead star, Aldebaran, looking extremely red. This star marks the Bull's eye. A little higher in the sky is the best known of all star clusters, the Pleiades, or Seven Sisters.

Looking South

The middle of the sky belongs to Pegasus (Flying Horse). Four of its stars form an unmistakable square. The square provides a useful guide for finding the misty patch that is the Andromeda Galaxy, which lies nearby.

The lower part of the southern sky is occupied by faint and "watery" constellations, such as Pisces (Fishes), Cetus (Whale), and Aquarius (Water-Bearer). The bright star appearing low in the sky near the horizon is Fomalhaut in Piscis Austrinus (Southern Fish).

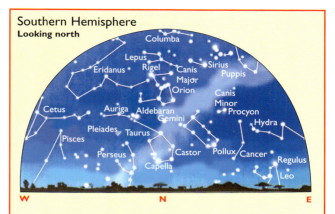

Southern Hemisphere
Looking north

October in the Southern Hemisphere means spring is well on its way. Pegasus and its famous square sits in mid-sky, almost due north. The upper part of the sky is less interesting, filled with faint "watery" constellations.

The Andromeda galaxy and the Pleiades cluster appear quite low in the sky. Toward the west, the bright pair of stars Altair (Eagle) and Deneb (Swan) are also low in the sky and will soon be setting.

Looking South

At this time of the year, the constellations in this part of the sky are faint and difficult to recognize. These include "watery" ones, such as the River Eridanus, and the "flock" of southern birds, such as Phoenix, Grus (Crane), and Pavo (Peacock).

Only close to the horizon does the sky brighten. Bright Rigel and Canopus are rising in the east, while Sagittarius (Archer) and Scorpius (Scorpion) dazzle in the west.

Southern Hemisphere
Looking south

Winter Stars

Winter brings chilly, frosty nights and clearer, darker skies. To the viewer looking north in mid-January, the Milky Way stands vertically. Only the tail end of Cygnus (Swan) is still visible, with Deneb shining brilliantly. Directly above, in mid-sky, is the unmistakable W-shape of Cassiopeia. Higher still, nearly overhead, is Capella, brightest star of Auriga (Charioteer) and the sixth-brightest in the whole heavens.

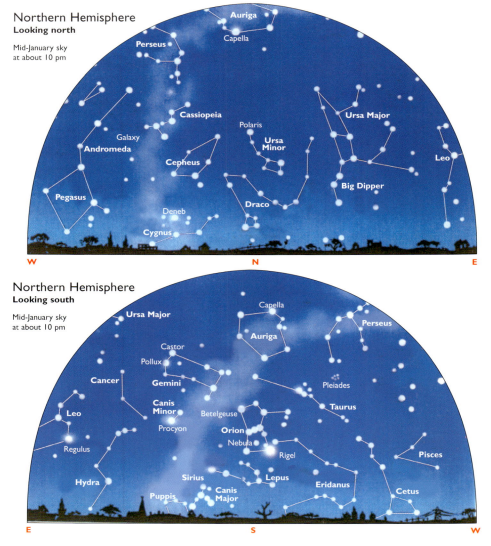

Northern Hemisphere
Looking north

Mid-January sky
at about 10 pm

Northern Hemisphere
Looking south

Mid-January sky
at about 10 pm

Looking South

This month's view of the southern sky is one of the finest of all. It is dominated by the most magnificent of all the constellations, Orion, but also boasts in mid-sky Taurus (Bull) and, across the Milky Way, Gemini (Twins).

Orion features two bright and contrasting stars, orange Betelgeuse and pure white Rigel. It also has a bright nebula, which can be seen with the naked eye under the three stars that form Orion's Belt. This nebula is a vast cloud of glowing gas and dust, where stars are being born all the time.

Orion is not only a brilliant constellation, but also a valuable sign-post to other highlights of winter skies. These highlights include the Pleiades star cluster and the brightest star in the heavens, Sirius. Sirius is also called the Dog Star because it is found in the constellation Canis Major (Great Dog).

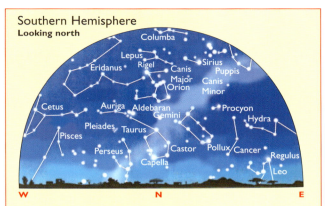

Southern Hemisphere
Looking north

In January, the Southern Hemisphere is in the middle of summer. The nights are short and relatively bright, so they are not ideal for stargazing. Looking north, Orion, Gemini, and Taurus delight the eye.

The brightest stars of these and other constellations form a dazzling ring. They are Rigel, Aldebaran, Capella, Castor, and Pollux, Procyon, and Sirius.

This is a good time to look at the Pleiades, also known as the Seven Sisters. However, you are unlikely to spot all of the seven brightest stars that gave this star cluster its name.

Looking South

Crux (Southern Cross) has just risen above the horizon in a vertical Milky Way. Above it come the brilliant "nautical" constellations of Vela (Sails), Carina (Keel), and Puppis. Carina's brightest star, Canopus, is second in brightness only to Sirius.

Southern Hemisphere
Looking south

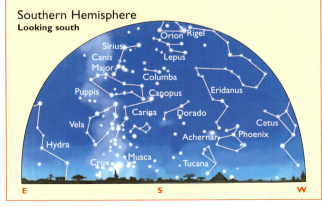

107

OBSERVING THE HEAVENS

Although we can see a lot of celestial objects in the night sky just with our eyes, we can see much more with the help of binoculars and telescopes. Astronomers use huge telescopes and also send instruments into space to spy on the strange and wonderful bodies that make up our universe.

At the simplest level, all you need to be an astronomer is your eyes. If you stargaze for any length of time, you will soon learn to recognize the patterns of bright stars in the sky — the constellations. You will see how they seem to wheel across the heavens during the night, and how they come and go with the seasons.

You will also notice the very bright stars that often appear to wander among the constellations. These are not stars but planets. They look bigger and brighter than the ordinary stars because they are much closer to earth — millions of miles away rather than trillions of miles.

The biggest and closest body of all, of course, is the Moon. The Moon is responsible for some of the most wonderful sights in nature — eclipses of the Sun — when day turns suddenly into night.

There are other unusual sights in the heavens that can be seen with the eyes alone. Almost every night you can see bright streaks where stars seem to be falling from the sky. Astronomers call these falling stars meteors.

From time to time, large objects with streaming tails appear, blaze through the heavens for a time, and then disappear. These spectacular bodies are comets, icy lumps that have travelled to our skies from the depths of the solar system.

There is, however, a limit to how much you can see with just your eyes. Astronomers therefore use instruments that gather much more light than our eyes can. They are telescopes, a word meaning something like "seeing far."

William Herschel, discoverer of the planet Uranus, built this enormous telescope in 1789, which had a body tube 40 feet (12 metres) long. He used it to discover Saturn's moons Mimas and Enceladus.

An observatory dome sits high above the clouds at a mountaintop observatory, where the air is crystal clear. At night the dome opens to expose its powerful telescope to the heavens..

Looking at Telescopes

The kind of telescope Galileo used to look at the heavens is known as a refractor, or refracting telescope. It is so called because it has glass lenses that refract, or bend light. Many amateur astronomers use refractors, which are easy to build in small sizes and easy to set up and use.

An amateur astronomer uses a small refractor. Such instruments are ideal for beginners.

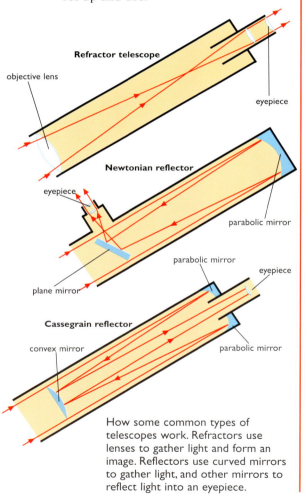

How some common types of telescopes work. Refractors use lenses to gather light and form an image. Reflectors use curved mirrors to gather light, and other mirrors to reflect light into an eyepiece.

Two main kinds of lenses are used in a refractor. At the front is the larger one, which is called the objective or object glass. The purpose of this lens is to gather incoming starlight and focus it — form it into a sharp image.

The other main lens is the eyepiece, which the astronomer looks through. The eyepiece is used to look at and magnify the image. It fits in a narrower tube than the objective and can be slid in and out for focusing.

Mirror Telescopes

About 60 years after Galileo had made his first observations, England's Isaac Newton developed a new kind of telescope, which used mirrors to gather light. This kind of instrument is called a reflector, or reflecting telescope, because mirrors reflect light. Most amateur astronomers who use reflectors have instruments called Newtonians, which are based on Newton's original design.

A Newtonian reflector has a shallow, dish-shaped mirror at the base of the telescope body tube. It gathers the incoming starlight and reflects the light to another plane (flat) mirror back up the tube. In turn, this mirror reflects the light into an eyepiece in the side of the tube.

Big Eyes

Generally speaking, the bigger a telescope is, the more it can see of the heavens, revealing even faint objects. The biggest telescopes in the world, the ones used by professional astronomers, are reflectors.

It is difficult to build really big refractors. The lenses have to be supported in the tube of the telescope so that light can pass through them. Big lenses are very heavy and cannot easily be supported in this way. The world's biggest refractor is at the Yerkes Observatory in Wisconsin, which has a 40-inch (1 metre) objective lens.

Because mirrors reflect light from their surface, they can easily be supported from underneath, no matter how big they are. The first giant reflector was the Hale, which has a mirror 200 inches (5 metres) across. Completed in 1948 at Mount Palomar Observatory in California, it is still one of the world's finest instruments.

Today, the most powerful reflectors have even bigger mirrors. They contain not a single flat mirror but one made up of many curved sections. Computers are used to help fit the sections in place to form a perfectly shaped surface. The two Keck telescopes in Hawaii have mirrors 33 feet (10 metres) across, made up of 36 separate sections.

Telescope domes at Mauna Kea Observatory on Hawaii. The photograph captures the moment of totality during the spectacular total eclipse of the Sun on July 11, 1991.

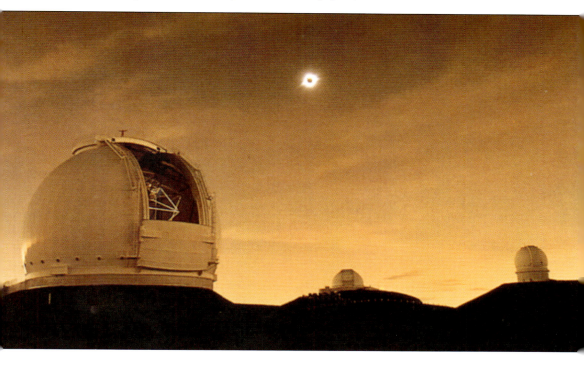

Astronomers at Work

Astronomers go to work at a time — night — when most other people are thinking of going to bed. They carry out their observations and study the results at observatories. The telescopes they use are housed in great domes, which provide protection from the weather. The roofs of the domes open at night to expose the telescopes to the heavens.

Most observatories are sited high on mountain peaks. There, they are above the thickest and dirtiest layers of the atmosphere. Mauna Kea Observatory in Hawaii is one of the highest, located at an altitude of more than 13,500 feet (4,200 metres).

Taking Pictures

These days professional astronomers do not just look through telescopes. Instead they use them as giant cameras and take photographs with them.

There is a very good reason for this. Photographic film can store the light that falls on it. The longer a film is exposed, the more light it will store. Astronomers often expose film in their telescopes for hours at a time. This allows the film to pick up the feeble light from the faintest of objects and make it visible.

When they take such long-exposure photographs, astronomers must allow for the fact that the stars move in relation to Earth. Otherwise the

Astronomers generally use black and white film to record images of the heavens. By exposing the film for long periods, they can pick up images of distant objects, such as this spiral galaxy in the constellation Centaurus, which lies millions of light years away.

Below: The distinctive McMath solar telescope at Kitt Peak National Observatory. Kitt Peak is one of the world's finest observatories, located on a mountain peak near Tucson, in Arizona.

photographs will be blurred. The telescopes are mounted on supports so that they move at the same speed and in the same direction as the stars.

Going Electronic

Most telescopes are now guided and controlled with the help of computers. Some are even controlled over long distances through radio links.

Electronics now also plays a vital role in making observations with telescopes. Many telescopes now record images of the heavens on electronic chips instead of photographic film. Called charge-coupled devices, or CCDs, they are similar to the chips used to form the image in videocameras. They are more sensitive to light than film.

Looking at Starlight

The most useful of the many other instruments astronomers use is the spectrograph. They use this instrument to study the faint light from stars in minute detail.

Starlight is like sunlight. It looks white but it is actually made up of a mixture of different colours. We see the colours in sunlight when we pass it through a prism. They form a spectrum. In a similar way a spectrograph splits starlight into a spectrum. By studying dark lines in the spectrum, astronomers can tell all kinds of things about a star, such as its temperature, what it's made of, and how fast it's travelling through space.

Above: An astronomer works with a 40-inch (1-metre) telescope at the Roque de los Muchachos Observatory in the Canary Islands. Like all modern instruments, it has an open body tube and is computer controlled.

Invisible Astronomy

Most astronomers learn about the heavens by studying the light that stars and galaxies give out. But studying starlight by itself does not give a true picture of what the universe is like. This is because stars give off not only light rays we can see but also many other kinds of rays that are invisible.

These rays include gamma rays, X-rays, and radio waves.

Only by looking at all the rays stars give out — visible and invisible — can astronomers build up a complete picture of what the universe is like. But there is a problem. Most invisible rays from space are partly or completely blocked by Earth's atmosphere. Only radio waves get through easily.

The Radio Window
The person who discovered this was a

This steerable radio telescope collects radio waves with its huge dish, which focuses them on the antenna above. The radio signals are processed by computer into false-colour images.

communications engineer named Karl Jansky, who worked at the Bell Telephone Laboratories in Holmdel, New Jersey. In 1931, he was investigating the hissing noises, or interference, that plagued long-distance communication by short-wave radio. He built a radio receiver from an antenna that circled round a track on the wheels from on old Model T Ford.

Jansky traced much of the interference to local sources or to the atmosphere. But some interference remained, no matter where he pointed the antenna. He suddenly realized that the interference was coming from the sky. The heavens were beaming down radio waves on the Earth.

Below: The Very Large Array radio telescope at Socorro, New Mexico. Its 27 dishes can be arranged in different patterns to concentrate incoming radio waves.

Jansky's discovery paved the way for a whole new branch of astronomy. Radio astronomy is now one of the most exciting branches of astronomy. It has led to the discovery of intriguing heavenly bodies such as quasars, pulsars, and immensely energetic radio galaxies.

Radio Telescopes

Astronomers gather the radio waves that come from the heavens with radio telescopes. They are quite unlike ordinary light telescopes. Most take the form of huge metal dishes, with an antenna in the centre.

The dish has to be huge to pick up the heavenly radio signals, which are very faint. It focuses the signals on the antenna, which then feeds them to a very sensitive radio receiver. Using computers, astronomers can then process the signals and display them as radio "pictures."

The largest dish radio telescope is built in a natural mountain valley on the Caribbean island of Puerto Rico. It measures 1,000 feet (305 metres) across. Some radio telescopes have several smaller dishes that work together to form a much bigger collecting area. A notable example of this type of radio telescope is the Very Large Array near Socorro, New Mexico. It uses 27 movable dishes, each 82 feet (25 metres) across.

Far right: Spacewalking astronauts visit the Hubble Space Telescope regularly to make repairs and update instruments.

Above: This Hubble picture shows delicate filaments of glowing gas around the stars of the Pleiades star cluster.

Space Astronomy

To study the other invisible rays given off by stars, astronomers use space technology. They send their telescopes and other instruments into space on spacecraft that travel well above Earth's atmosphere.

Some spacecraft are satellites, which circle Earth repeatedly in orbit. Others are probes, which escape from Earth completely and travel deep into the solar system to visit planets and their moons, asteroids, and comets.

Astronomy Satellites

Satellites have been launched to study all the invisible radiation given off by stars — gamma rays, X-rays, ultra-violet rays, infrared rays, and microwaves. The results they have sent back have added greatly to our knowledge of the universe and how it began and developed.

The most outstanding astronomy satellite looks at the universe in visible light. It is the Hubble Space Telescope (HST), launched from the space shuttle orbiter Discovery in April 1990.

The HST was expected to look much farther into space than ever before and show stars and galaxies in fantastic detail. But at first, the pictures it sent back were not much better than the best Earth-based telescopes. The curve of its light-gathering mirror was slightly off because of a problem with the way it had been made.

However, a daring space mission in December 1993 repaired the HST, and it began beaming the most fantastic pictures back to Earth. They showed stars being born in spectacular clouds; solar systems in the making; stars puffing off great masses of gas as they die; the remains of stars that have blasted themselves apart; and huge donuts of glowing matter that could hide the most awesome of heavenly objects — black holes.

Space Probes

Space probes have been flying into deep space since the 1960s. They have visited seven of the eight planets besides Earth. Only the most distant planet, icy Pluto, has not yet been explored from space.

The planets and the other heavenly targeted by these probes are extremely far away. Probes take months to reach even the closest planets, Venus and Mars. They take years to reach the others. Probes take so long because for most of the journey time they coast, or travel without power. It would be impossible to provide them with enough fuel to power them all the way.

Sometimes, to reduce the time of their journey, probes loop round other planets and use their gravity to increase speed. This technique is called gravity-assist. It was used by the Voyager probes on their spectacular missions to Jupiter, Saturn, and Uranus, Galileo also used gravity-assist to reach Jupiter, as did Cassini to reach Saturn.

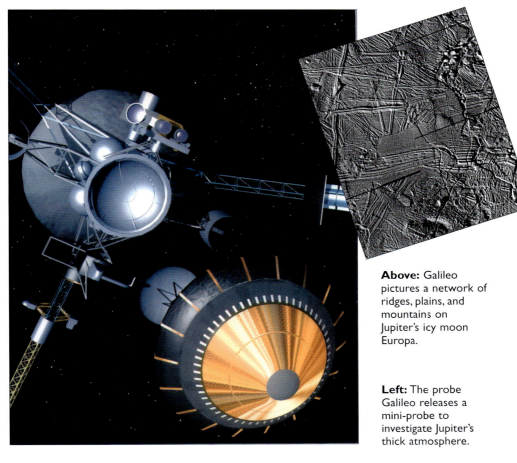

Above: Galileo pictures a network of ridges, plains, and mountains on Jupiter's icy moon Europa.

Left: The probe Galileo releases a mini-probe to investigate Jupiter's thick atmosphere.

Exploring the Sun and Moon

The Sun and the Moon are the two heavenly bodies that dominate our skies — the Sun by day and the Moon by night. When the Moon is full, they both appear about the same size in the sky. This does not mean that they are really the same size, of course. In reality, the Sun is 400 times wider across (in *diameter*) than the Moon. It looks to us to be the same size as the Moon only because it is 400 times farther away.

The Sun and the Moon are two very different kinds of bodies. The Sun is a great ball of searing hot gases that pours out enormous amounts of light and heat into space. The Moon is a drab ball of cold rock that shines only because it reflects the Sun's light.

And there are still more differences. The Moon circles around Earth, once a month. Earth circles around the Sun once a year. Earth circles around, or orbits, the Sun along with eight other bodies, the planets. Earth itself is a planet. The planets are the main members of the solar system. (*Solar* means having to do with the Sun.)

In our corner of space, the Moon is not particularly important at all. There are many other moons. The Moon is not essential to life on Earth. But there is only one Sun. Without its light and heat, Earth would be a dark, cold, and dead world, without life as we know it.

In part because it is so much closer to Earth, we know more about the Moon than we do about the Sun. Astronomers have been studying the Moon for hundreds of years, with their eyes and through telescopes. More recently, space probes have explored it, and astronauts have roamed across its surface and brought back samples of rocks and soil. It is likely that one day soon humans will return to the Moon and set up permanent bases there.

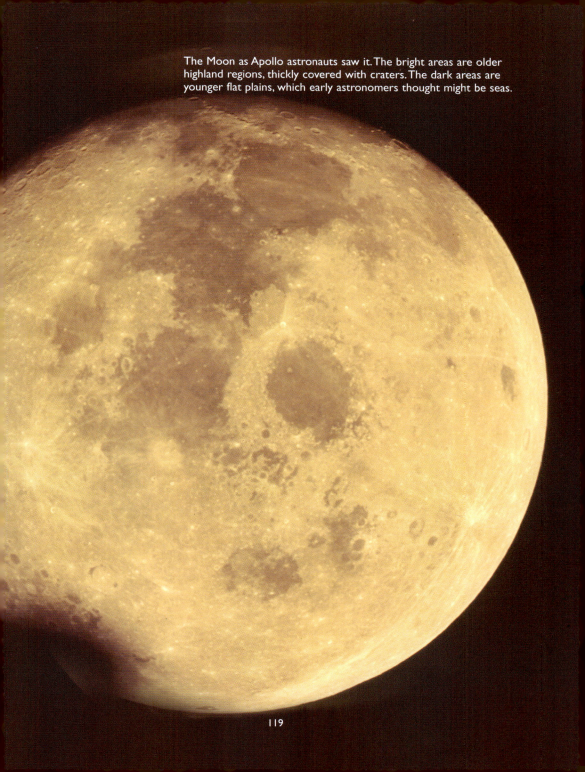

The Moon as Apollo astronauts saw it. The bright areas are older highland regions, thickly covered with craters. The dark areas are younger flat plains, which early astronomers thought might be seas.

OUR STAR, THE SUN

The Sun is Earth's local star. It looks much bigger than the other stars we see in the heavens only because it is much closer. The Sun is very special to us, but it is not special in the universe. Indeed, it is a very ordinary star — there are billions like it.

The Sun is about 93 million miles (150 million km) away from Earth. This may seem a great distance, but in space it is just a short step. The next nearest star, called Proxima Centauri, lies more than 25 trillion miles (40 trillion km) away. The light from this star takes more than 4 years to reach Earth. Sunlight takes only about 8½ minutes to reach Earth.

Compared with Earth, the Sun is enormous. It could swallow more than a million bodies the size of Earth. It measures about 865,000 miles (1,400,000 km) in diameter.

Among stars, the Sun is much bigger than some but much smaller than others. Huge stars called supergiants are hundreds of times bigger in diameter, which is why astronomers classify the Sun as a dwarf star. They call it a yellow dwarf because it gives off yellowish light.

To people on Earth, it appears that the Sun moves across the sky every day. But it is not the Sun that is moving but Earth. Earth spins around in space on its axis, which makes it seem that the Sun is moving through the sky.

But the Sun does move in two ways. One, it spins around on its axis — or revolves — just like Earth does. The Sun, however takes about 25 days to revolve once. Two, the Sun is travelling through space at a fantastic speed — about 12 miles (20 km) per second. The billions of other stars in our local star system, or galaxy, are travelling at similar speeds.

Like other stars, the Sun was born in a nebula, a great cloud of gas and dust that once existed in our corner of space. Astronomers estimate that this happened about 4.6 billion years ago. The cloud shrunk into a ball, which became hotter and hotter until it began to shine as a star. The rest of the Sun's family of planets and other bodies formed at the same time.

Opposite: The fiery ball of the Sun, pictured by the SOHO probe. Bright flares are erupting in several places, while a huge looping fountain of flaming gas is shooting thousands of miles above the surface (top right).

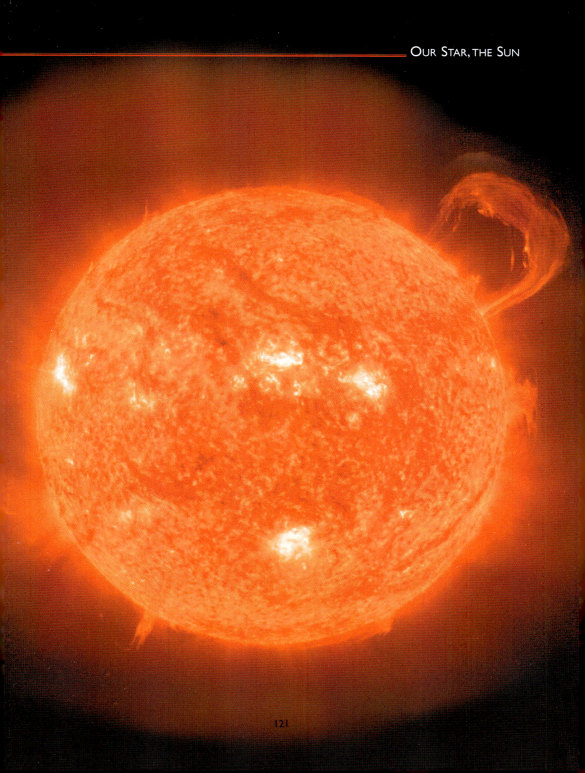

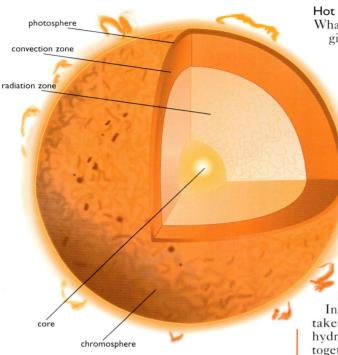

photosphere

convection zone

radiation zone

core

chromosphere

Hot Stuff

What is the source of the heat and light given off by the Sun? It cannot be produced by burning an ordinary fuel such as natural gas, or methane. If the Sun burned such a fuel, it would have burned itself out billions of years ago.

The Sun produces its energy by nuclear re-actions. These are reactions in which atoms of elements take part. Atoms are among the tiniest particles of substances that can be measured. The centre of an atom is called the nucleus, which is where the term "nuclear" comes from.

Fusing Atoms

In the main nuclear reaction that takes place in the Sun, atoms of hydrogen are made to fuse, or join, together. When this happens, fantastic amounts of energy are given off in the form of light and heat. This kind of reaction takes place in the centre, or core, of the Sun, where temperatures reach 15,000,000°C or more.

From the core, the energy travels outward towards the Sun's surface. It may take as long as 100,000 years to reach the surface. First it travels as radiation (waves), then on moving gas currents — rather like the hot air currents you can feel when you put your hand over a radiator at home.

When the energy does reach the surface, it escapes into space as light, heat, and other radiation. The Sun's surface is called the photosphere, meaning the light-sphere. It is very much cooler than the core, with a temperature of about 5,500°C.

Inside the Sun

Like the other stars, the Sun is a great ball of gas. Actually it is a mixture of many gases. Astronomers learn which gases are present in the Sun by examining the light it gives out. They have learned that the most common gas is hydrogen, which is the lightest gas of all. The next most common gas is helium.

Along with hydrogen and helium, astronomers know that the Sun contains many other substances, such as iron, calcium, and sodium. These substances are solids on Earth but are gases on the Sun because it is so hot.

Carry on Shining

The nuclear reactions that occur in the Sun use up huge amounts of hydrogen — some 600 million tons every second. But the Sun is so big that it has enough hydrogen to last for about another 5 billion years.

When all the hydrogen is used up, the Sun will start to die. First, it will swell up to become a kind of star called a red giant. Then it will shrink until it becomes a white dwarf — a tiny, very dense body only about the size of Earth.

When the Sun dies, it will first expand into a red giant before shrinking again. Then it will throw off masses of gas to form a planetary nebula rather like the Ring Nebula (above). Over time, all that will be left is a tiny hot ball of matter — a white dwarf.

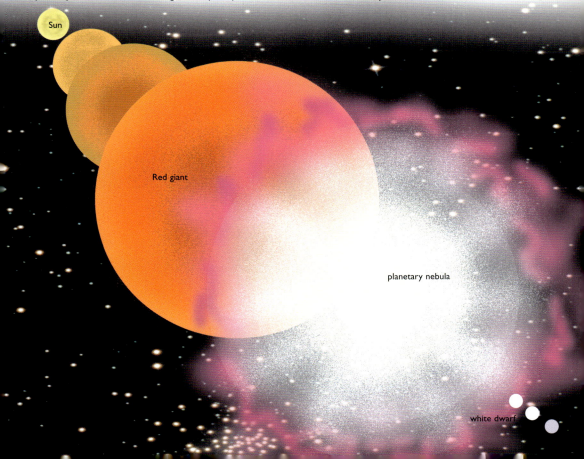

Sun

Red giant

planetary nebula

white dwarf

Because the photosphere is so blindingly bright, we usually cannot see the chromosphere or corona. We can see them only during a total eclipse of the Sun when the Moon blots out the Sun's glare.

During an eclipse, the chromosphere can be seen as a pink ring around the dark Moon. The corona is much more spectacular, billowing out from the Sun and

Above: The Sun's surface has a speckled, or grainy, appearance, which astronomers call granulation. The three bright patches in the picture are violent explosions called flares.

The Solar Atmosphere

The photosphere forms the outer layer of the Sun. It is about 200 miles (320 km) thick. But the Sun does not end at the photosphere's outer layer because it has an atmosphere, or layer of gases, around it, just as the Earth does.

The lower part of the Sun's atmosphere is called the chromosphere, or colour sphere. This is because it has a pinkish colour. A thinner outer atmosphere begins about 3,000 miles (5,000 km) above the surface. It is known as the corona, which means "crown." It extends into space for millions of miles.

Above: A huge tongue of flame leaps above the Sun's surface. It is a fountain of hot gas known as a prominence. In a matter of hours, prominences can climb 100,000 miles into space.

stretching pearly white into space for a vast distance.

The Stormy Surface

The photosphere looks glaringly bright all over. But closer inspection — by special instruments — shows that its surface is actually a patchwork of tiny bright and dark areas. They look rather like grains of wheat, and astronomers call them granules. Each granule is a little rising column of gas that carries heat from below to the surface.

Larger bright and dark regions constantly appear and disappear on the photosphere. The most obvious are dark ones that look rather like ink blots. They are known as sunspots. Sunspots are regions that are about 1,500°C cooler than their surroundings. Sunspots come and go in a predictable way over a period of 11 years. This period is known as the sunspot cycle.

Fountains and Flares

Sunspots are caused by changes in the Sun's magnetism. They trigger off other spectacular happenings, such as prominences. These great fountain-like streams of glowing gas may leap above the Sun for hundreds of thousand of miles. They often form into loops, following invisible lines of magnetism.

Sometimes, around new sunspots, there are enormous explosions called flares. For a few minutes, a flare can become the brightest thing on the Sun. It gives off not only light but also invisible radiation, such as X-rays and radio waves. Flares also give off atomic particles that flow out into space.

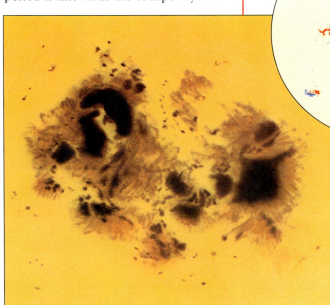

Above: This picture highlights regions of intense magnetism on the Sun's surface. They are found around sunspots.

Left: A typical sunspot group, showing darker (umbra) and lighter (penumbra) regions. Sunspot groups can spread over tens of thousands of miles.

125

The Sun's Family

Earth is one of nine planets that circle in space around the Sun. These planets are the main bodies in the solar system. They are scattered at great distances from one another. Most of the solar system just consists of empty space.

The Earth and its three neighbours in space — Mercury, Venus, and Mars — are quite close together in the inner part of the solar system. The planets in the outer solar system — Jupiter, Saturn, Uranus, Neptune, and Pluto — are much farther apart.

The diagram shows the orbits of the planets drawn to scale. Notice that the orbits of all the planets except Pluto are in much the same plane. This means that they would lie on or close to an imaginary flat sheet in space. But Pluto travels quite far above and below it.

Size Matters

The diagram also shows the great difference in sizes among the planets. The four inner planets are tiny in comparison with the next four huge outer planets. They are also quite different in composition, or what they are made of.

The four inner planets are made up mainly of rock, while the four large outer planets are made up mainly of gas and liquid gas. The farthest planet, tiny Pluto, is different again, being made up of rock and ice.

Other Bits and Pieces

Besides the planets, there are many other smaller bodies in the solar system. Most planets have one or more satellites, or bodies, circling them. Earth has only one — the Moon — but Saturn and Uranus each have at least 18.

Another major group of bodies are the asteroids. They are found in a wide band, or belt, roughly midway between the orbits of Mars and Jupiter. Even the biggest asteroid, Ceres, is only about 600 miles (950 km) in diameter. Most of the thousands of asteroids discovered by astronomers are much smaller.

The comets that occasionally appear in our skies are small icy bodies that journey in towards the Sun from the most distant regions of the solar system.

The space between the planets also contains even smaller bits of rock and metal. These bits rain down on Earth all the time, burning up as they hit the upper air. We see them as fiery streaks called meteors, which are also popularly known as falling or shooting stars.

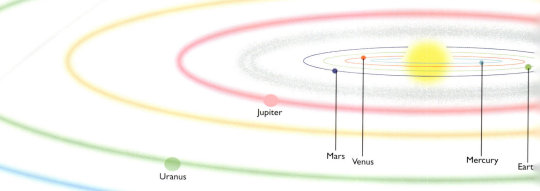

Jupiter

Mars Venus Mercury Eart

Uranus

Neptune

Pulling Power

All these bodies in the solar system — planets, moons, asteroids, comets, and little chunks of matter — are all held together by the Sun's gravity. The Sun has some 750 times more mass that all the other bodies in the solar system put together. This gives it an enormously powerful gravitational force — so powerful that it can attract icy comets only a few miles in width that are located trillions of miles away.

Below: The orbits of the nine planets around the Sun, drawn roughly to scale. They occupy a region of space some 7 billion miles (12 billion km) across.

Jupiter

Saturn

Uranus

Pluto

Neptune

Earth

Venus

Mars

Mercury

Pluto

SUN AND EARTH

Life on Earth could not exist without the Sun. Its light and heat brighten and warm our world and make it a suitable home for human beings and millions of other living things. It helps determine the weather and creates our climate. It helps us measure time.

Even prehistoric peoples realized how important the Sun was. Many early civilizations worshipped the Sun as a god. More than 4,000 years ago, the ancient Egyptians worshipped the Sun god Ra and pictured him sailing across the heavens during the day. More recently, in Central and South America, the Inca, Aztec, and Mayan peoples centreed their religion on Sun worship. They practiced human sacrifice on a vast scale to insure that the Sun rose every day.

Until only about 550 years ago, most people thought that the Sun travelled around the Earth. It certainly appears to every day. In the 15th century, a Polish churchman and astronomer named Nicolaus Copernicus suggested that Earth orbited the Sun. This marked the beginning of the concept of the sun-centred, or solar, system. Science has since proven that Copernicus was right.

Earth is located in just the right place in the solar system for the creation and sustenance of life. It receives just the right amount of heat from the Sun to keep it at a comfortable temperature for plant and animal life.

If Earth were closer to the Sun, it would be too hot for liquid water — for life. Look at the planet Venus, which is closer to the Sun — it is hotter than an oven. On the other hand, if Earth were much farther away from the Sun, it would be too cold for liquid water — and for life as we know it. For example, Mars, which is farther away from the Sun, has no life on it.

Earth teems with life in great variety. The tiger is one of the several million different species (kinds) of animals and plants found on Earth.

"Spaceship Earth," photographed by
Apollo astronauts. The overall colour of
Earth is blue because this is the colour
of the great oceans that cover more
than two-thirds of the surface.

Left: The ancient Egyptians thought that the god Ra carried the Sun across the heavens every day in a boat.

Sun Time

To people on Earth, the Sun seems to travel across the sky every day. It rises above the horizon in the east in the early morning, then travels westward, seeming to climb all the while. It reaches its highest point in the sky at noon, or midday. In the afternoon, still travelling westward, it gradually sinks lower in the sky. In the evening, it sets beneath the horizon in the west.

The time between when the Sun reaches its highest point in the sky one day and the time it reaches its highest point the next day is always the same. This is the period of time we call a day. The day is one of our basic units for measuring time. We split this natural division of time artificially into 24 hours, each hour into 60 minutes, and each minute into 60 seconds.

In a Spin

As Copernicus pointed out centuries ago, the Sun does not actually travel around Earth. It only seems to, and that is because Earth itself spins around once every day. Earth spins around on its axis in space from west to

Left: The astronomical clock at Hampton Court, in London, England. It shows not only the time of day but such things as the phases of the Moon, the time of the year, and the constellation of the zodiac.

east, and this makes the Sun appear to travel in the opposite direction — from east to west.

The Solar Year

Earth travels around the Sun in a nearly circular path. It spins around 365 and one-quarter times while it travels one complete circle, or orbit. In other words, it takes Earth 365¼ days to orbit the Sun once. This is another basic unit of time, which we call the year.

We use the day and the year to make up our calendar — a standard way of dividing up time. As well as the day and the year, our calendar also uses months, a time period loosely based on the time it takes the Moon to complete its phases.

Our calendar year is 365 days, except for every fourth year, when an extra day is added at the end of February, the shortest month in the calendar year. This extra day is added to account for the difference of one-quarter day

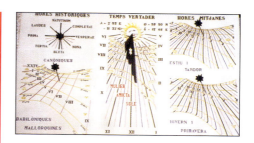

An unusual, modern Spanish sundial, designed to be accurate during different seasons of the year.

between the calendar year and the natural, or solar year.

Leap years are usually years that can be divided by four, or by 400 for beginning-of-the century years, such as 2000. Even so this system does not keep our calendar year exactly in time with the solar year. So sometimes extra seconds — leap seconds — are added to keep our calendar in step with the workings of the heavens.

Earth spins around on its axis once a day, and it takes 365¼ days, or 1 year, to travel once in its orbit around the Sun. The day and the year are the two basic units by which we measure time.

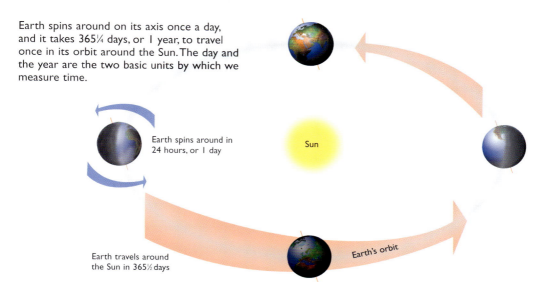

Earth spins around in 24 hours, or 1 day

Sun

Earth's orbit

Earth travels around the Sun in 365½ days

The Weather Machine

The Sun pours out enormous amounts of energy into space. And only a tiny fraction of it — one two-billionth — reaches Earth. But even this tiny amount is enough to warm the Earth and make its weather.

Beams of sunlight

equator

The warmth, or temperature, of our surroundings is the most important feature of Earth's weather. It affects all the other features, too, such as the movement of air and the amount of moisture in the air.

For example, air rises over hot regions, creating areas of low pressure. It sinks over cold ones, creating areas of high pressure. Air masses flow from high to low pressure, creating winds and even larger movements of the air.

The Sun heats the water in oceans and lakes, and makes it evaporate, or escape into the air as vapour (gas). The water vapour rises into the air and cools down. When it cools sufficiently, it turns back into droplets of water, which collect to form clouds. When the drops

get big enough, they fall back to the ground as rain or even snow if the temperature is low enough.

Snow and the Sun

Of course, not all places on Earth have the same weather. For example, in winter, many people from snowy New England go on vacation to Florida, where it is sunnier and much warmer.

In general, throughout the year Florida has warmer weather than New England — it has a warmer climate.

Florida has a warmer climate than New England because it is nearer to Earth's Equator. Places near the Equator enjoy

- 🔴 Tropical
- 🩷 Subtropical
- 🟡 Desert/steppe
- 🟠 Mediterranean
- Warm Temperate
- Cold Temperate
- Mountain
- Polar

the warmest climate on Earth because throughout the entire year they receive the most direct sunlight.

Because the Earth is round, places farther away from the Equator receive less direct sunlight. To people living there, the Sun never appears to climb as high in the sky as it does to people at the Equator. Accordingly, those places get less direct sunlight. The farther a place is away from the Equator, the lower the Sun will climb and the cooler it will be. Places near the North and South Poles are coldest of all.

Although the climate of a place depends mainly on its distance from the Equator, it can also be affected by other things. For example, places on sea coasts enjoy a warmer climate than places in the middle of the continents. This is because water retains heat better than land does. Ocean currents affect climate, too. For example, parts of northwestern Europe enjoy a mild climate because the Gulf Stream helps warm them.

World Climates

Scientists divide the world into a number of regions that have the same kind of climate. They enjoy similar temperatures and have similar rainfall throughout the year. The map shows the world divided into eight kinds of climate, or climatic, zones. Each zone has its own kind of vegetation and wildlife, which can thrive in that particular climate.

Climates around the world. Hot, dry deserts cover huge expanses of Earth's surface. Warm temperate regions are productive regions for agriculture. Vast conifer forests thrive in the northern cool temperate regions.

New England regularly has snow in winter, while Florida enjoys the sunshine and a warmer climate.

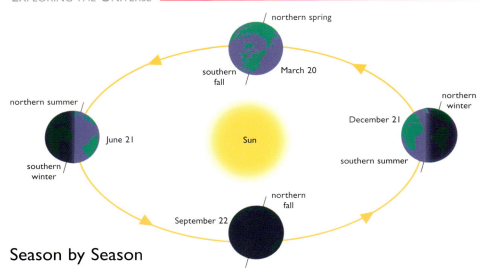

northern spring

March 20

southern fall

northern summer

June 21

Sun

northern winter

December 21

southern winter

southern summer

northern fall

September 22

Season by Season

Generally, the weather experienced in a particular place depends not just on its location on Earth but on the time of year. For example, much of the United States, and the rest of the northern hemisphere, is cold in December and days are short. By March the weather is warming up and the days are getting longer. June sees hot weather and long days. By September, temperatures are cooler and the days get shorter again. The same pattern is repeated every year. It is a natural rhythm of nature. In the southern hemisphere, the seasons are the direct opposite of what they are in the north. When it is winter in the northern hemisphere, it is summer in the southern hemisphere, and so on.

The regular changes in temperature and the lengths of the day and night divide the year into periods we call the seasons. Over much of the world there are four seasons — winter, spring, summer, and fall (or autumn).

In some parts of the world, there are not four distinct seasons. Near the Equator, the weather hardly changes at all as the months go by. On either side of the Equator, in the tropics, there are only two seasons, a dry one and a rainy one.

Earth's Tilt
What causes the changing seasons? It is the way Earth moves in space in relation to the Sun. As Earth travels in its yearlong orbit around the Sun, it spins around like a top on its axis, which

Opposite: Earth's tilt in space brings about the seasons. The northern hemisphere is tilted most toward the Sun on about June 21, and tilted farthest away from the Sun on about December 21.

is an imaginary line through its centre and the North and South Poles.

But Earth does not spin upright as it circles the Sun. Its axis is tilted (at an angle of 23½ degrees). It remains tilted at this angle all the time and always points in the same direction in space.

This means that as Earth travels in its orbit, the axis points first towards, then away from, the Sun. So different places on Earth are tilted more towards the Sun at some times than they are other times. The more they are tilted towards the Sun, the warmer they will be. The more they are tilted away from the Sun, the cooler they will be. This is what brings about the change of temperature with the seasons.

Changing Skies

You can follow the changing seasons during the day by watching how high the Sun climbs in the sky at midday. It climbs highest in summer and lowest in winter. We can also follow the seasons by watching the night sky.

Earth circles around the Sun during the year. On any particular date, we look out at night in a particular direction in space. We see certain star groups, or constellations. As the days and months

go by and Earth moves in its orbit, we look into a slightly different spot in space, so we see different constellations.

For example, in North America, the constellation Orion appears in the southern part of the evening sky in the winter. But in the fall, the constellation Pegasus appears in that part of the sky.

Opposite: "Down under" in Australia in December, the people are enjoying summer. They often celebrate Christmas by having parties on the beach.

Right: In the tropical rain forests that grow near the Equator, the climate stays the same all year: it is always hot and wet.

THE SILVERY MOON

The Moon is Earth's closest neighbour in space and its constant companion. As it circles around Earth once a month, it appears to change shape, from a slim crescent to a full circle, then back again. The pull of the Moon's gravity causes the daily rise and fall of the oceans that we call the tides.

The Moon is Earth's only natural satellite. It is quite a small body, with a diameter of 2,160 miles (3,476 km). This is about one-quarter the diameter of Earth, and nearly three-quarters the distance across the United States.

Although the Moon is relatively small, it dominates the night sky because it is so close to Earth. On average, it is only about 239,000 miles (384,000 km) away. The next nearest body to us in space — Venus — is more than 100 times farther away. It never comes closer to Earth than 26,000,000 miles (42,000,000 km).

Other planets besides Earth also have moons, of course. Among them, Earth's moon ranks fifth in size. Bigger than the Moon are the three largest of Jupiter's moons — Ganymede, Callisto, and Io — and Saturn's moon Titan. However, Earth's moon is much bigger in relation to the size of its parent planet than are these other moons. Astronomers sometimes refer to Earth and the Moon together as a double planet.

Long ago, people worshipped the Moon. Just as the Sun brought light to the world by day, so the silvery Moon brought light by night. The ancient Romans called their Moon goddess Diana. In Latin, the language spoken by the ancient Romans, the word for Moon was "luna," and from this we get our word "lunar," meaning to do with the Moon.

Some ancient peoples also believed that the Moon somehow affected people's minds, particularly the full moon. They believed that exposure to moonlight could cause a person to go mad. This led to the term "lunatic" being used for an insane person.

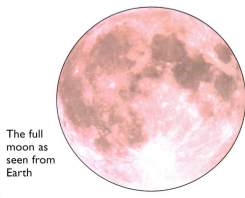

The full moon as seen from Earth

This astronaut's view of a full moon looks different from the full moon
we see from Earth. The right-hand side of the photograph reveals part of
the far side of the Moon, which we can never see from Earth.

Sun's rays

last quarter

Earth

new moon

full moon

first quarter

new moon

first quarter

full moon

last quarter

As the Moon circles around Earth during the month (left), we see different amounts of its face lit up by the Sun at different times (right). These views are the main phases.

Moon Motions

The Moon gives off no light of its own. The Moon appears to shine because it reflects light given off by the Sun. The Moon does not reflect sunlight very well — it reflects less than ten percent of the light that falls on it. Astronomers say it has a low albedo ("whiteness").

As the moon circles the Earth, we see different amounts of it lit up at different times each month. This accounts for what we refer to as the "phases" of the moon.

Monthly Motion

The Moon circles once around Earth every 27⅓ days. During this time, Earth completes part of its yearlong orbit around the Sun. It takes 29½ days for the Moon to complete its phases — to go through all the stages of illumination in which it appears in the sky.

This lunar period is one of the great natural divisions of time. Early peoples used this lunar month as the basis of

their calendars. Today, we base our calendar on the solar year and use months of different numbers of days so that they fit into this year.

Waxing and Waning

The actual shape, or phase, of the Moon we can see at any time depends on where the Moon, the Sun, and Earth are located in space. Once a month, the Moon moves roughly in line between the Sun and Earth in space. Then we can't see it in the night sky because the Sun lights up only its hidden far side. The nearside that faces us remains dark, in shadow. We call this phase of the Moon the new moon.

A day or so after new moon, the Sun lights up the edge of the nearside, and we see the Moon as a crescent, or bow shape. A week later, the Moon, the Sun,

A werewolf is on the prowl! The idea of the werewolf, a man who turns into a wolf at the time of the full moon, is a favourite theme in horror movies.

Left: The full moon phase, which we see once a month. One of its most prominent features is the bright ray system around the crater Tycho (lower centre).

Below: The thin sliver of the crescent moon, when the Moon is 27 days old. Soon the Moon will disappear completely, at the next new moon.

and Earth are again roughly lined up in space. The Moon is now on the opposite side of Earth from the Sun. And so the Sun lights up all the nearside of the Moon in the night sky.

We call this phase the first quarter. A week later the Sun lights up all the nearside, and we call this phase the full moon. During the time the Moon has been growing from a crescent to full, we say that is has been waxing. After full moon, the amount of the nearside lit up by the Sun gets smaller as time goes by. A week after full moon, only half is lit up. We call this phase the last quarter. And a week later only a thin crescent can be seen. Then the Moon disappears completely — it is the next new moon. During the time the Moon has been shrinking in size, we say it has been waning.

139

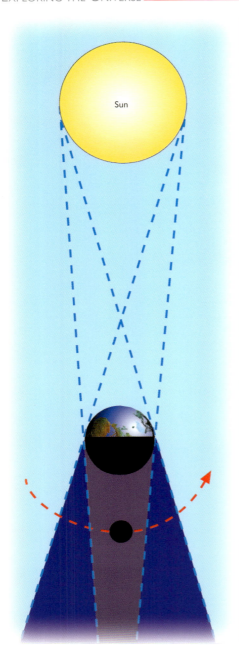

Captured Rotation

From Earth, we always see the same side of the Moon. That is because the Moon rotates once around in space in 27⅓ days, which is exactly the same amount of time it takes to orbit Earth. Several other moons have the same kind of motion, spinning around once in the same amount of time it takes them to orbit their home planet. Astronomers call this a captured rotation. It seems to be a natural rhythm of the universe.

Into the Shadows

The Moon travels in a more or less circular orbit around Earth. And Earth travels in a more or less circular orbit around the Sun. Twice a month, the Sun, Earth, and Moon line up roughly in space — at the time of full moon and new moon.

They only line up roughly because the Sun, the Moon, and Earth do not travel in quite the same plane, or flat sheet, in space.

But just a few times a year, the three bodies line up exactly in space. At times, Earth moves exactly into line between the Sun and the Moon, and its shadow falls on the Moon. We call this

an eclipse of the Moon or a lunar eclipse. At other times the Moon moves exactly into line between the Sun and Earth, and its shadow falls on Earth. This is called an eclipse of the Sun or a solar eclipse.

Eclipse of the Sun

The most spectacular eclipses are eclipses of the Sun. They take place at a new moon. Because the Moon is a small body, it casts a small shadow, which only ever covers a small area of Earth's surface.

If you are in this area during an eclipse, you will first see the Moon edging across the face of the Sun. The light will gradually start to fade. Then, when the Moon covers up the face of the Sun completely, the sky will get dark. Night will fall — during the day. We call this a total eclipse.

But total darkness will last for only a short time — about 7 minutes at most. Then the Moon moves on, the Sun reappears, and day returns once more.

Sometimes the Moon doesn't quite cover up the Sun — we call this a partial eclipse. Sometimes a ring of light remains around the edge of the Moon — we call this an annular (ring) eclipse. On average, two or three solar eclipses of one kind or another take place each year somewhere in the world.

Far left: During an eclipse of the Moon, the Moon moves into the Earth's shadow in space. It can stay in eclipse for more than two hours.

Opposite: During a lunar eclipse, the Moon does not disappear from view. It turns a pinkish colour as it is lit up by light coming from Earth's atmosphere.

Three stages of a solar eclipse:

1. The Moon is gradually covering up the face of the Sun.

2. The Sun's face is now completely covered. The Sun's outer atmosphere, or corona, appears brilliant white.

3. The Moon is just uncovering the edge of the Sun, and the total eclipse is over. This stage is sometimes called the "diamond ring."

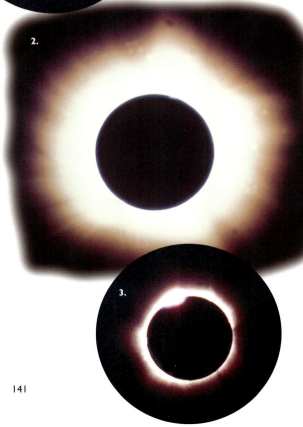

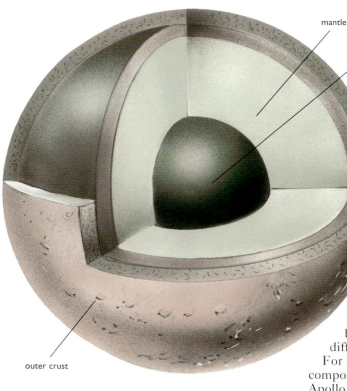

mantle

core

outer crust

Left: The Moon is made up of several layers. Its hard outer crust is up to about 45 miles (70 km) thick. Then comes a deep layer of rock known as the mantle. At the centre is a large core, which may contain metals such as iron.

What the Moon is Like

Astronomers sometimes refer to Earth and the Moon as a double planet. They are both rocky bodies that lie close together in space, but in most respects they are very different.

For one thing, their chemical composition is different. The rocks the Apollo astronauts brought back from the Moon are much different, chemically, from rocks on Earth.

Below: The Moon has weak gravity and no atmosphere. Astronauts on the Moon have to wear spacesuits in order to breathe.

The Moon's structure is also different from Earth's. Scientists have worked out the structure by studying "moonquakes," or ground tremors on the Moon. They received information about these from seismometers left behind by the Apollo astronauts.

Weak Gravity

The Moon is much smaller than Earth and therefore has much less mass. This also means that it has a very low gravitational force, or pull. In fact, on the surface it has only one-sixth the gravitational pull of Earth.

The low gravity has largely dictated what the Moon is like. With such low gravity, the Moon also has little atmosphere. Without an atmosphere, there is no weather like we have on Earth — no winds, clouds, rain, or snow. The sky is not blue but black. Without an atmosphere, the Moon is a silent world, because sound needs air to travel through.

Without a "blanket" of air, there is a great difference in temperature on the Moon's surface between day and night. During the day, temperatures rise to more than 100°C — the boiling point of water. During the night, they fall to 150°C below freezing (0°C). A "day" on the moon lasts two weeks on Earth.

Making the Tides

The Moon feels the effect of Earth's gravity. It is Earth's pull that keeps the Moon circling Earth once a month. In its turn, the Moon's gravity — weak though it is — affects Earth. In particular, it causes the tides, the regular rise and fall of the oceans.

The Moon tugs on the ocean waters beneath it, causing a high tide. It also tugs on Earth and pulls it away from the waters on the opposite side of Earth. In between the high tides, the water falls to create low tides.

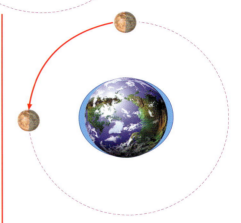

When the Moon is overhead, it exerts its strongest gravitational pull on the part of Earth beneath it. Because water is liquid, it is pulled more easily than the land and therefore rises. The result is that water accumulates on the parts of Earth directly beneath and directly opposite the position of the Moon. This is called a high tide, and it occurs twice a day at any given place. The Sun's gravity also affects the tides. The greatest range between high and low tides occur at the new moon, when the Moon and Sun are closest together in the sky, and at the full moon when they are directly opposite.

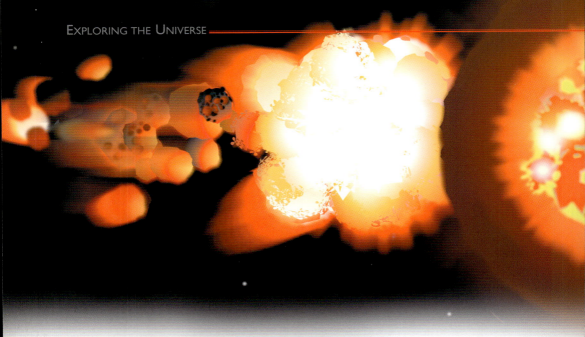

Birth of the Moon

When the Apollo astronauts returned from the Moon, they brought back many rock samples. Scientists found that the oldest rocks were more than 4 billion years old. The Moon itself must have formed before this, either at the same time as Earth and all the other planets (4.6 billion years ago), or soon after.

No one knows exactly how the Moon was formed. There are several theories. One is that the Moon formed at the same time as Earth. Another is that the Moon was once part of Earth. The newly formed Earth was molten (liquid) and spinning around rapidly. Eventually, it flung off into space a huge blob of molten matter, which cooled and became the Moon.

The trouble with these two ideas is that neither of them explains the differences in chemical make-up and structure between the Moon and Earth.

A theory that does explain these differences is that the Moon was formed in another part of the solar system from different materials. Sometime long ago, it strayed near Earth and was captured by Earth's gravitational pull. However, no one has been able to give a good explanation of how this happened.

Impact

The most convincing explanation of how the Moon was formed suggests that it was created as the result of a collision between another large body and the newborn Earth. The force of this body — which could have been as big as Mars — smashing into Earth ripped off a big lump of Earth matter. This matter mixed with matter from the colliding body and became the Moon.

Cooling Down

However the Moon formed, it would

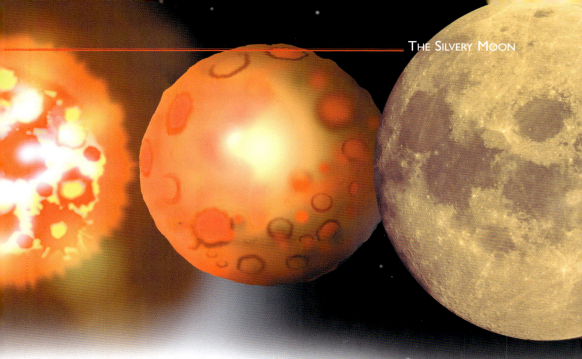

first have been in a molten state. Then it slowly cooled, and a wrinkled skin, or crust, formed over it. At this time, space was full of great chunks of rock that had not yet formed into planets or moons. These chunks rained down on the young Moon and dug great craters in its surface hundreds of miles across. This bombardment lasted for hundreds of millions of years.

While the Moon's crust cooled, the rocks inside were heating up. This happened because of radioactivity. Radiation substances in the rocks gave off radiation and also heat. Over time they melted and oozed out as lava into the great pits dug by the chunks of rocks from outer space. The lava spread out to form vast flat plains.

These are the plains referred to today as the moon's "seas." Many are still surrounded by circular mountain ranges that were created as the result of the original collisions.

Most of the craters found on the Moon were created during the fierce bombardment of the surface that happened billions of years ago.

THE LUNAR SURFACE

The Moon's surface has remained more or less as it is today for nearly three billion years. It displays a multitude of fascinating features — vast dusty plains, soaring mountain ranges, and craters galore.

Some features of the Moon's surface can be seen even with the naked eye. Two main regions can be detected — light and dark. The Italian astronomer Galileo first used a telescope to look at the Moon in 1609. He observed that the dark areas were flat regions and the light areas were highlands.

Comparing the Moon's surface with Earth's, he called the flat regions maria, or seas, and the highlands terrae, or continents. As more powerful telescopes were developed, it became apparent that the flat regions were not watery seas, but dry plains. Even so, the term "seas" is still used to refer to these lunar regions.

There are craters all over the Moon. There are fewer of them, however, on the seas than in the highlands. This tells scientists that the seas are much younger than the highlands. Astronomers think that the highlands are part of the Moon's original crust.

As well as craters, the Moon has many other distinctive features. These include domes, which are swellings of the crust made by lava pushing up. There are long trenches that snake across the surface, sometime for 100 miles (160 km) or more. Called rills, they probably formed when channels or tubes carrying lava collapsed. Most of these features are typical of the lunar seas.

The highest parts of the highlands are the mountain ranges surrounding the seas. They rise as high as 20,000 feet (6,000 metres).

The Apollo astronauts explored both sea and highland regions when they explored the Moon on foot in the 1960s and 1970s. On their six landing missions, they roamed the surface for more than 80 hours and brought back 850 pounds (385 kg) of Moon rocks. They also took thousands of stunning photographs. These expeditions remain the only human exploration of another heavenly body.

Opposite: The lunar surface has been shaped by rocky bombardment from outer space and also by volcanic processes going on in the Moon's interior. The landscape is barren, with a beauty all its own.

Moonwatch

The picture opposite shows the Moon, as we see it once a month, at the time of the full moon. Some of the main seas, mountains, and craters are marked. If you cover the left-hand side with a sheet of paper, what remains visible is the Moon's eastern hemisphere (half). This is how the Moon appears at its first quarter phase, when it is about 7 days old (7 days after new moon).

The Eastern Half

Several seas are found in the eastern hemisphere (half) of the Moon. Four of them merge into one another — the seas of Serenity, Tranquillity, Fertility, and Nectar. The first Apollo landing took place on the Sea of Tranquillity on July 20, 1969. The easiest sea to spot lies near the eastern limb (edge). It is the small Sea of Crises, which is about 300 miles (500 km) across.

The high ranges of the Apennines and the Caucasus Mountains separate the Sea of Serenity from the Sea of Showers to the west. Peaks in these ranges soar to 20,000 feet (6,000 metres)

or more. In the north, a prominent valley, called Alpine Valley, links the Sea of Showers with the Sea of Cold.

Of the craters in this hemisphere, Aristoteles and Eudoxus are prominent in the north. A line of large craters runs south from the centre, beginning with the 92-mile (150-km) wide Ptolemaeus. These craters are best seen at the first quarter phase, when they show up on the terminator — the boundary between the lit and unlit parts of the Moon.

The Western Half

If you cover the right-hand side of the picture opposite, you will see the Moon's western hemisphere. This is how you will see the Moon at its last quarter phase, when it is about 21 days old.

This half of the Moon is dominated by two huge seas, the circular Sea of Showers and the sprawling Ocean of Storms. The Sea of Showers is the largest of the circular seas, measuring more than 700 miles (1,100 km) across. It merges into the Ocean of Storms, which extends over an area of over 2,000,000 square miles (5,000,000 sq km), or about two-thirds the size of the United States.

On and around the edge of the Sea of Showers are prominent craters, such as the dark-floored Plato, Archimedes, and Copernicus. Copernicus and nearby Kepler show up brilliantly at full moon, when they are ringed by bright rays. But even they are outshone by Tycho, whose brilliant rays stretch for hundreds of miles.

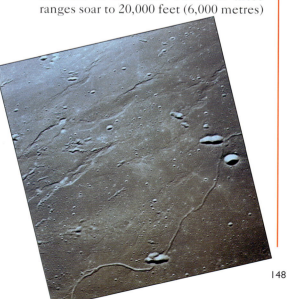

The surface of the Sea of Tranquillity. Like the other seas, it is relatively flat and smooth. Little ridges and channels snake across the surface. They were made by lava flows pushing up under the surface, or flowing over it.

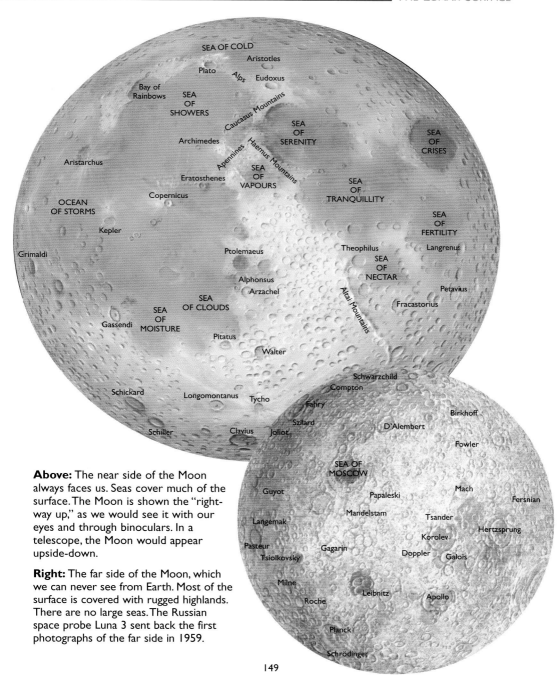

SEA OF COLD
Aristotles
Plato
Alps
Eudoxus
Bay of
Rainbows
SEA
OF
SHOWERS
Caucasus Mountains
SEA
OF
SERENITY
SEA
OF
CRISES
Archimedes
Apennines
Haemus Mountains
Aristarchus
Eratosthenes
SEA
OF
VAPOURS
SEA
OF
TRANQUILLITY
Copernicus
OCEAN
OF STORMS
Kepler
SEA
OF
FERTILITY
Grimaldi
Ptolemaeus
Theophilus
Langrenus
SEA
OF
NECTAR
Alphonsus
Arzachel
Petavius
Fracastorius
SEA
OF
MOISTURE
SEA
OF CLOUDS
Altai Mountains
Gassendi
Pitatus
Walter
Schickard
Longomontanus
Tycho
Schwarzchild
Compton
Fahry
Birkhoff
Szilard
D'Alembert
Schiller
Clavius
Joliot
Fowler
SEA OF
MOSCOW
Guyot
Papaleski
Mach
Fersnian
Mandelstam
Langemak
Tsander
Hertzsprung
Pasteur
Korolev
Gagarin
Doppler
Galois
Tsiolkovsky
Milne
Leibnitz
Apollo
Roche
Planck
Schrödinger

Above: The near side of the Moon always faces us. Seas cover much of the surface. The Moon is shown the "right-way up," as we would see it with our eyes and through binoculars. In a telescope, the Moon would appear upside-down.

Right: The far side of the Moon, which we can never see from Earth. Most of the surface is covered with rugged highlands. There are no large seas. The Russian space probe Luna 3 sent back the first photographs of the far side in 1959.

Craters

Craters are by far the most common feature of the lunar surface. They are found in the seas and the highland regions. Almost all of the craters were made by meteorites colliding with the Moon. A few are volcanic craters, meaning that they are the openings of volcanoes that erupted long ago.

Craters come in all sizes, from shallow dents in the ground to depressions more than 100 miles (160 km) across. The biggest one that can be easily seen on the Moon's nearside is called Clavius. It measures some 145 miles (260 km) across.

Most large craters have walls higher than the surrounding area and floors that are lower. The floor of a crater called Newton is more than 5.5 miles (9 km) deep. Many large craters also have another feature — a central mountain range.

Craters Old and New

Over time, craters themselves might be hit by meteorites. In the process they can become damaged, or "ruined." Their walls might be knocked down, or new craters are formed within the old one. Sometimes craters become so filled with lava that little of them remains. These are often called ghost craters.

Some of the large newer craters show up brilliantly at full moon because of their rays — lines of shiny material leading from them like the rays of the Sun. The rays consist of material thrown out when the crater was formed.

Right: Craters cover nearly the whole of the far side of the Moon. This heavily cratered region is thought to be part of the Moon's original crust, little changed for billions of years.

Below: A stunning view of the crater Copernicus, taken by an orbiting space probe. Note the step-like terraced walls and the central mountain range, which are typical of the large lunar craters.

Moon Rocks

We know a lot about Moon rocks because of the samples collected by the Apollo astronauts. In general, they are similar to some rocks found on our own planet. They are all igneous ("fire-formed"), which means they formed when molten rock cooled. Many were formed when volcanoes erupted on the Moon billions of years ago.

There are no sedimentary rocks on the Moon as there are on Earth. Sedimentary rocks form from layers of sediment — material that settles out of rivers and seas. Of course, there has never been any great amount of water on the Moon, let alone rivers or seas.

Basalts and Breccias

There are two common kinds of Moon rocks. One is dark-coloured and made up of tiny crystals. It is like the rock known as basalt on Earth.

The other main type is a mixture of rock chips cemented together. It was formed when meteorites hit the surface, melting some rocks and shattering others. The molten rock flowed over bits of broken rock and cemented them together. This type of rock is called breccia.

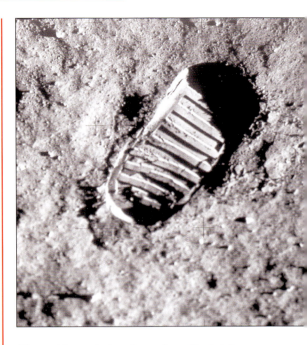

Above: Human beings from planet Earth left their footprints in the lunar soil, which is known as regolith. The soil is a mixture of fine dust, rock chips, and minute glass beads.

Below: A typical light-coloured Moon rock, like some volcanic rocks we find on Earth.

Project Apollo

To send astronauts to the Moon and bring them back safely to Earth required an extraordinary effort on the part of thousands of people who worked with the U.S. government organization known as NASA (National Aeronautics and Space Administration). For the Apollo project, which was the name given to the Moon-landing effort, a new generation of spacecraft and rockets had to be developed. Astronauts had to be trained in the techniques needed to bring about a lunar landing and a safe return. This took time and money — about $25 billion by the time the project had finished.

Lunar Orbit Rendezvous

The method NASA developed to land astronauts on the Moon was called lunar orbit rendezvous. It centred on the three-man Apollo spacecraft. The whole spacecraft would set out for the Moon. Then, in lunar orbit, a landing vehicle would separate from the main "mother ship" and descend to the lunar surface. Later, it would take off from the surface and rendezvous (meet) with the mother ship, which would then return to Earth.

The Apollo spacecraft was made up of three modules (parts). The crew occupied the command module, which

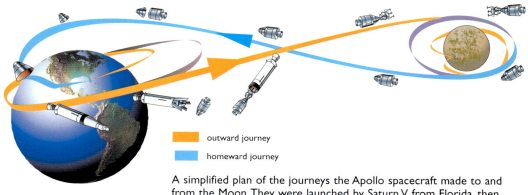

outward journey

homeward journey

A simplified plan of the journeys the Apollo spacecraft made to and from the Moon. They were launched by Saturn V from Florida, then separated and flew to the Moon. The lunar module dropped down to the surface, then carried its two astronauts back to the mother ship, which returned to Earth. Just before re-entering the atmosphere, the command module separated and splashed down in the Pacific Ocean.

Left: The three-man Apollo spacecraft, shown as it neared the Moon, with the lunar module still linked with the command module.

was pressurized with oxygen for the crew to breathe. This was joined for most of the time with an equipment, or service, module, which had a powerful rocket engine. The command and service modules together formed the CSM mother ship.

For the trip to the Moon, the CSM was linked to the landing vehicle, the lunar module (LM). This was built to take two of the astronauts down to the surface. They lifted off the Moon in the upper part of the LM, using the lower part as a launch pad.

Mighty Moon Rocket

Rocket pioneer Wernher von Braun was the mastermind behind the powerful rocket needed to launch the 45-ton Apollo spacecraft to the Moon. This rocket, the Saturn V, was gigantic. With the Apollo spacecraft on top, it stood 365 feet (111 metres) tall and weighed nearly 3,000 tons.

To launch the Saturn V, a new launch site was built just inland from Cape Canaveral in Florida. It was named the Kennedy Space Centre after John F. Kennedy, the president who launched the Apollo Project. The site featured two launch pads and a huge Vehicle Assembly Building (VAB), where the Moon rockets were put together. Today, the Kennedy Space Centre is the focus for space shuttle operations.

153

Men on the Moon

A Saturn V first carried astronauts to the Moon in December 1968. It was a test mission (Apollo 8), which orbited the Moon on Christmas Eve and returned without landing. Television viewers back on planet Earth were able to see the surface of the Moon in close-up for the first time. What a sight it was!

The following July, Apollo 11 set off to attempt the first Moon landing. On July 20, 1969, the Apollo 11 lunar module (code-named Eagle) touched down on the Sea of Tranquillity. Reported lunar module commander Neil Armstrong to mission control at Houston: *"Tranquillity Base here. The Eagle has landed."*

A few hours later Armstrong became the first man to set foot on the Moon. *"That's one small step for man,"* he said, *"one giant leap for mankind."* And so it was. A being from planet Earth had set foot on another world.

Moonwalking

Armstrong was the first of 12 astronauts to leave footprints in the lunar soil. Over the next three-and-a-half years, they explored the lunar seas and highlands, first on foot and then with the help of a lunar rover, nicknamed the "Moon buggy."

Opposite, main picture:
The most famous Apollo picture, showing Edwin Aldrin on the Sea of Tranquillity during the Apollo 11 mission in July 1969. **Inset:** The Apollo 11 crew, from the left: Neil Armstrong, Michael Collins, and Edwin "Buzz" Aldrin.

Right, main picture: On the last three Apollo missions the astronauts were able to wander farther afield because they had transportation — the lunar rover, or Moon buggy. **Inset:** All the Apollo missions splashed down in the Pacific Ocean, lowered gently into the sea by three huge parachutes.

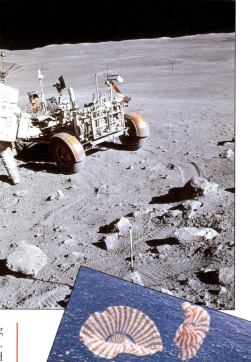

The astronauts had a punishing workload, planned in detail beforehand. They picked up rocks and also drilled into the ground for samples. They also set up scientific stations, using a package of different instruments called ALSEP (Apollo lunar surface experiments package).

A nuclear battery powered the equipment, which included a seismometer to detect tremors in the ground, or "moonquakes." The ALSEP scientific stations radioed the data they collected back to scientists on Earth and continued to do so until 1977.

"We Shall Return"

Apollo 17 commander Eugene Cernan took the last step on the Moon on December 14, 1972. Just before he left, he made this promise: "We leave as we came and, God willing, we shall return with peace and hope for all mankind."

No doubt human beings will one day return to the Moon. This time they will probably build permanent residences and set up scientific bases, observatories, and mining camps as well. The recent discovery of water (as ice) in some lunar craters means that future bases could be more self-supporting.

Exploring the Near Planets

The Sun rushes headlong through space the solar system in tow. Earth and eight bodies called planets form the major part of the solar system. Each planet circles around the Sun at varying distances from it. In order of distance from the Sun, the nine planets are Mercury, Venus, Earth, Mars, Jupiter, Saturn, Uranus, Neptune, and Pluto.

The four planets nearest the Sun — Mercury, Venus, Earth, and Mars — lie quite close together in space. Often referred to as the inner or near planets, they have much in common. They are very different from the planets farthest from the Sun — Jupiter, Saturn, Uranus, Neptune, and Pluto — which are referred to as the outer or far planets.

The common link between all the near planets is that they are made mainly of rock. That is why they are called

An artist's impression of the Pathfinder probe and the Sojourner rover on Mars. In the background are the mountains known as the Twin Peaks.

terrestrial, or Earth-like, planets. Although they are similar in composition, they are different in size and have developed along different lines since their birth some 4.6 billion years ago.

Mercury is the smallest of the inner planets and the one closest to the Sun. It is a scorching world scarred with craters. Temperatures can rise to 840°F (450°C). Venus is the hottest planet with the temperature as high as 900°F (480°C). Its volcanic landscape is hidden from view beneath permanent clouds. Earth is a green and pleasant land thanks to its comfortable temperatures and the presence of liquid water. It is the only planet on which conditions are suitable for life as we know it. Mars is much smaller, and much cooler than Earth.

All the inner planets are visible in the night sky, and their existence has been known since people began gazing at the stars. Venus is the most unmistakable of the planets, hanging like a brilliant star in the west just after sunset on many nights of the year. It is often referred to as the evening star. Mars is distinctive, too, recognizable by its fiery red-orange colour.

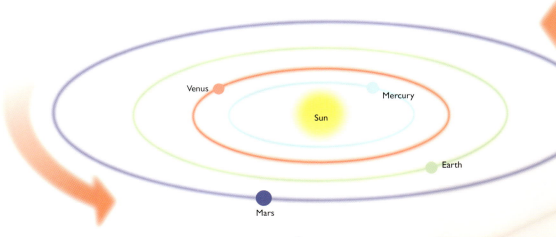

Paths of the Planets

The four inner planets — Mercury, Venus, Earth, and Mars — are located in the centre of the solar system. They are relatively close to one another, sometimes passing within a few tens of millions of miles. Venus, for example, sometimes approaches within 26 million miles (42 million kilometres) of Earth.

"Tens of millions of miles" might not seem close, but it is in terms of distances in space. Beyond Mars, space is fairly empty for hundreds of millions of miles until Jupiter. This giant body is the first of the widely scattered outer planets.

The planets circle around the Sun in regular paths, or orbits. The farther away they are from the Sun, the greater the distance they have to travel in their orbit, and the longer is their "year" — the time they take to circle the Sun. Mercury's year lasts only about 88 Earth-days, but Mars's year is nearly eight times longer.

In the Same Direction

As they orbit the Sun, the planets do not move in all directions. They all travel in much the same plane, or flat

Above: The four inner planets circle the Sun quite close together. Mars lies on average about 142 million miles (228 million km) away from the Sun, and occasionally comes within 35 million miles (56 million km) of Earth.

sheet, in space, and they all circle around the Sun in the same direction — counter clockwise, or in the opposite direction of the way the hands of a clock move.

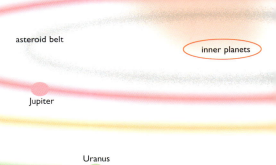

Circling in Ovals

Strictly speaking, it is not correct to say that planets "circle" the Sun. The orbits of the planets are an oval, or elliptical, path around the Sun. This means that the planets are closer to the Sun at some points in their orbits than they are at others. Among the inner planets, Mercury and Mars have the most oval, or eccentric, orbits. Mars's distance from the Sun varies between 129 and 155 million miles (207 and 249 million kilometres).

Planet Data

Planet	Av. distance from Sun in million miles (km)	Diameter at Equator in miles (km)	Completes orbit (length of year)	Rotates around in (length of day)	Mass (Earth=1)	Density (water=1)	Number of moons
Mercury	36 (58)	3,032 (4,880)	88 days	58.6 days	0.06	5.4	0
Venus	67 (108)	7,521 (12,104)	225 days	243 days	0.8	5.2	0
Earth	93 (150)	7,927 (12,756)	365.25 days	23.94 hours	1	5.5	1
Mars	142 (228)	4,221 (6,792)	687 days	24.63 hours	0.1	3.9	2
Sun	—	865,000 (1,392,000)	—	25 days	330,000	1.4	—

orbit of Pluto

Saturn

The five outer planets are more widely scattered through space. Even the nearest, Jupiter, never comes closer than about 460 million miles (740 million km) to the Sun. And the farthest planet, Pluto, wanders more than 4,500 million miles (7,300 million km) away.

Neptune

Size and Structure

Earth is the largest of the four inner planets. Venus is only slightly smaller. Then comes Mars and finally Mercury, which is only about one-third the size of Earth in diameter. Despite the differences in size between the four planets, astronomers believe that they all have a similar structure, or make-up.

Of course, we only know for certain what Earth is like inside. Geologists — the scientists who study Earth — have found this out by studying the way earthquake waves travel through the rocks. The waves suddenly change direction at different depths underground, indicating that they are passing through different layers.

The geologists have found that Earth is made up of two main layers of rock above a huge ball of molten metal, mainly iron and nickel. This ball forms the core, or centre, of Earth. The rock layers above it are the hard crust that forms Earth's surface and a denser layer of rock underneath called the mantle.

The inner planets, drawn to the same scale. The cutaways show how they are made up of different layers.

Looking at Layers

The other inner planets almost certainly have similar layers. But as the cutaway diagrams show, the size of the layers and the core varies widely.

Mercury seems to have only a thin crust and mantle but a huge metal core. Venus probably has a slightly smaller core than Earth but a thicker crust. Earth's core is different from those of the other planets because the outer part is liquid, in the form of molten metal. Scientists believe that this is the reason Earth has such a strong magnetic field in comparison with the other inner planets.

Mars is less dense than all the other inner planets. This indicates that it does not have as much metal in its core.

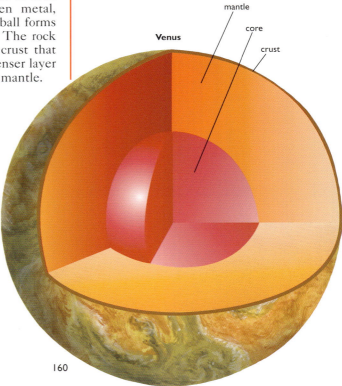

Venus

mantle

core

crust

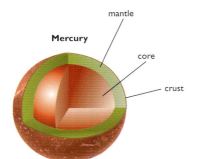

Mercury

mantle

core

crust

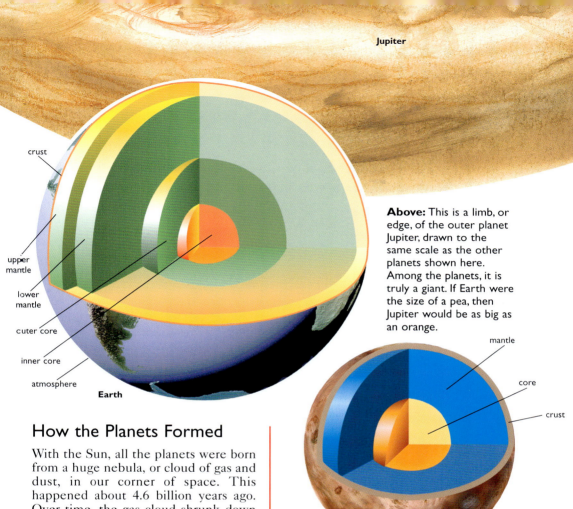

Jupiter

crust

upper mantle

lower mantle

outer core

inner core

atmosphere

Earth

Above: This is a limb, or edge, of the outer planet Jupiter, drawn to the same scale as the other planets shown here. Among the planets, it is truly a giant. If Earth were the size of a pea, then Jupiter would be as big as an orange.

mantle

core

crust

Mars

How the Planets Formed

With the Sun, all the planets were born from a huge nebula, or cloud of gas and dust, in our corner of space. This happened about 4.6 billion years ago. Over time, the gas cloud shrunk down into a hot ball (which would become the Sun), surrounded by a disk of matter.

The disk was made up of chunks of rock, metal, gas, and dust. In time the chunks came together to form larger and larger bodies, which gathered around them gas and dust. These were the infant planets.

Blasts of gas and particles from the Sun blew away the gas and dust around the four young inner planets into the outer part of the solar system, leaving them bare rock.

Over time, gases escaped from the rocky surface of the planets through volcanoes to form new atmospheres. Mercury was too small and too hot to hold on to an atmosphere. Venus and Earth were big enough and had strong enough gravity to keep a thick atmosphere. Being much smaller, Mars could hold on only to a much thinner atmosphere.

SPEEDY MERCURY

Mercury is the closest planet to the Sun. It is also the fastest-moving planet, hurtling along in its orbit at a speed of nearly 30 miles (50 km) a second. Appropriately, it is named after the fleet-footed messenger of the gods in Roman mythology.

Like Venus and Mars, Mercury can be seen from Earth with the naked eye. At best, it looks like a bright star. But it is the most difficult of the inner planets to spot. This is because it never strays far from the Sun in the sky and cannot be seen in really dark conditions.

Mercury can sometimes be seen as a morning star down low near the horizon in the east just before sunrise. In the northern hemisphere, fall is usually the best time to see it in the morning skies. Sometimes Mercury can be seen low in the western sky just after sunset as an evening star. Spring is usually the best time to see it in the evening sky.

Mercury can be seen more easily using binoculars or a telescope. Seen through a telescope, Mercury seems to change size over time. This happens as Mercury approaches Earth and then moves farther away as it makes its orbit around the Sun.

To observers on Earth, Mercury also seems to change shape over time. This is because we see different amounts of it lit up by the Sun at different times as it makes its orbit. We call these changing shapes its phases. They are like the changing shapes, or phases, of the Moon that we see every month.

Telescopes, however, are not powerful enough to show any details of Mercury's surface, except a few vague markings. Astronomers had no idea of what the surface was like until the space probe Mariner 10 flew past it in 1974. The planet turned out to be almost completely covered in craters, like the ancient highland areas of our Moon.

Left: Craters large and small dominate the surface of Mercury. The largest have central mountain peaks, just like the large lunar craters.

162

A montage of images sent back by the probe Mariner 10. It flew past Mercury three times — twice in 1974 and once in 1975.

The Long Day

Earth takes 24 hours to spin around once on its axis. This period of time is a day. But Mercury spins around much more slowly, taking nearly 59 Earth-days to spin around just once. It takes only 88 Earth-days to travel once around the Sun — this is Mercury's year.

This slow rotation and short "year" give Mercury a very long daytime (when it is in the sunlight) and a very long nighttime (when it is in darkness).

On Earth, there is only one day between one midday and the next. But on Mercury there are 176 Earth-days — in other words a Mercury "day" is 176 Earth-days long. Both daytime and nighttime last for about half this time — 88 Earth-days.

Scorching Hot

We would expect Mercury to be a hot planet because it orbits so close to the Sun. During Mercury's long daytime, places are baked by the Sun for 88 Earth-days at a time. Temperatures there soar as high as 430°C, or more than four times the temperature of boiling water. This is hot enough to melt lead.

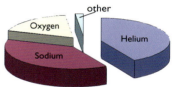

Make-up of Mercury's incredibly faint atmosphere.

MERCURY DATA

Diameter: 3,032 miles (4,880 km)

Average distance from Sun: 36,000,000 miles (58,000,000 km)

Mass (Earth=1): 0.06

Density (water=1): 5.4

Spins on axis in: 58.6 days

Circles round Sun in: 88 days

Number of moons: 0

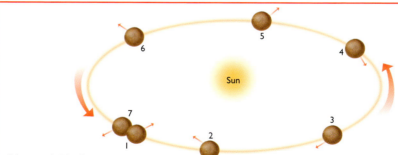

Mercury's Motion
The diagram shows how Mercury slowly spins around, as it orbits, or rotates around, the Sun. In position 1, the arrow on Mercury points directly toward the Sun. We can call this midday on Mercury. As the planet travels in its orbit, the arrow points in different directions. In 1 Mercury "year" of 88 days (7), it has spun round 1½ times, and the arrow is now pointing directly away from the Sun into the darkness. We can call this midnight. It will take another circle around the Sun (another Mercury "year") before it is again pointing toward the Sun. This will be the next midday. In other words, one "day" on Mercury, from midday to midday, lasts two Mercury "years," or 176 Earth-days.

But on the dark, night side of Mercury, it is a different story. Places stay in the dark for 88 Earth days at a time. During this period, they lose most of their heat to space. Temperatures fall as low as −180°C. This is the same temperature as the planet Saturn, billions of miles away from the Sun.

Airless World

Because Mercury is such a small planet, it has only a weak gravitational pull, which means it has not been able to hold on to a thick atmosphere like Venus's or Earth's. In any case, such an atmosphere would have been driven off ages ago by the intense heat of the Sun.

Nevertheless, astronomers have detected very slight traces of gas around Mercury. This gas constitutes a very, very thin atmosphere. It contains hydrogen and helium gases that have come from the Sun and atoms of sodium that have come out of the baked rocks.

An Iron Planet

Mercury could be called the iron planet because it contains more iron for its size than any other planet. Its iron core makes up as much as four-fifths of the bulk of the planet. Only about one-fifth is rock.

This huge mass of iron makes Mercury magnetic, just as Earth's iron core makes it magnetic. But Mercury's magnetism is about 100 times weaker than Earth's, so it is doubtful whether you could use a magnetic compass to find your way around the planet, as you can on Earth.

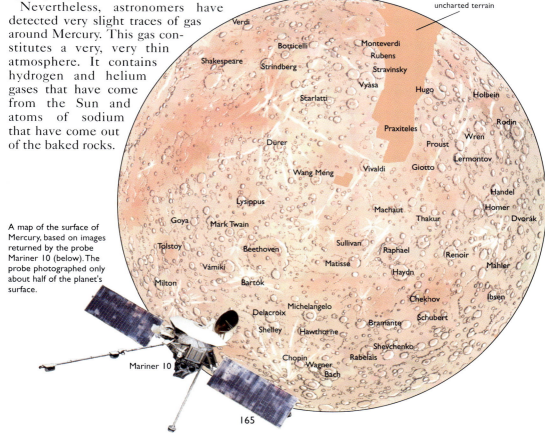

A map of the surface of Mercury, based on images returned by the probe Mariner 10 (below). The probe photographed only about half of the planet's surface.

165

Mercury Landscapes

When it flew past Mercury three times in 1974 and 1975, the space probe Mariner 10 was able to take photographs of the planet. These pictures revealed that the half of Mercury that could be seen from Mariner 10 is heavily cratered. There is no reason to think that the other half is any different.

At first sight, Mercury's surface looks like that of the Moon. But it doesn't have any large plains, or seas, as the Moon has. The craters, however, are just like those on the Moon. There are thousands upon thousands of them, ranging in size from shallow pits a few hundred feet across to large structures hundreds of miles in diameter. The large craters have raised walls, deep floors, and central mountain peaks.

The largest crater, named Beethoven, measures more than 400 miles (600 km) across, twice as big as the craters Shakespeare and Goethe. Most of the craters on Mercury are named after people famous in the arts, such as composers (Beethoven), playwrights (Shakespeare), poets (Goethe), and painters.

As mentioned already, there are no large plains, or seas, on Mercury, but there are smaller flat regions between many of the heavily cratered areas. These are known as intercrater plains.

Under Attack

Most of the craters on Mercury were created when it was bombarded by meteorites, or rocks from outer space. This happened mostly between about 4 and 4.5 billion years ago, not long after the planet formed.

Most of Mercury's craters date from this time. Many have since been flooded with lava, leaving them with

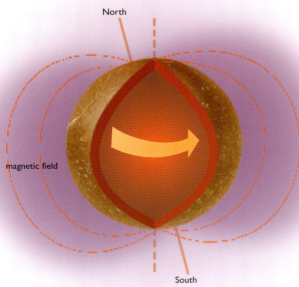

North

magnetic field

South

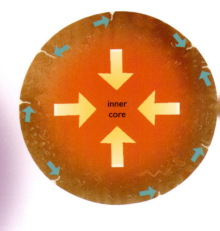

inner core

Mercury's huge iron core has had an enormous effect on the planet. It created a magnetic field (left). When the core shrunk a long-time ago, it caused the surface rocks to crack and pile up, creating steep cliffs.

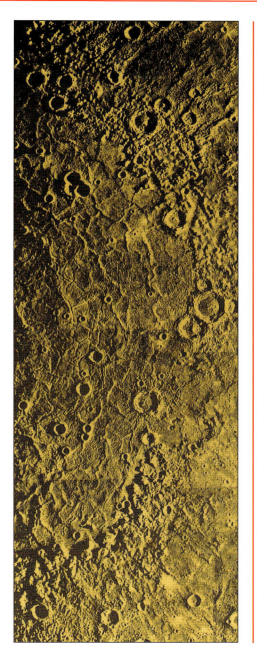

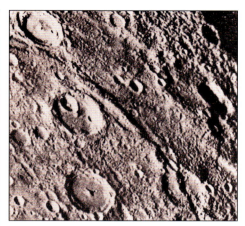

Above: Craters and cliffs scar Mercury's surface.

Left: Rings of mountains surround Mercury's Caloris Basin.

flat, smooth floors. Some craters are much younger. Bright lines lead out from them like rays from the Sun. That is why these lines are called crater rays. They are lines of shiny material thrown out when the craters were formed.

The Big Basin
Sometime after the main meteor bombardment of Mercury ceased, the planet was struck by a massive asteroid. It created a huge crater, over 800 miles (1,300 km) across. The force of impact melted the surface rocks, which flooded the crater with lava. The impact also sent out shock waves, which rippled across the surface and threw up the rocks into circular mountain ranges, a mile or so high.

This huge, mountain-ringed formation is called the Caloris Basin. It was named Caloris (meaning heat) because it lies close to one of the hottest spots on the planet.

167

VEILED VENUS

Venus comes closer to Earth than any other planet. It looks beautiful when it hangs in the sky as an evening star, and the Romans named it after their goddess of beauty and love. But up close, Venus is far from beautiful. It has a suffocating, crushing atmosphere with acid clouds.

Venus is the easiest of the planets to spot in the sky. For several months each year, it can be seen in the darkening western sky just after sunset. That is why it is often referred to as the evening star. It comes out long before the real stars.

At other times of the year, early risers can spot Venus in the brightening eastern sky just before sunrise. All the real stars set long before it.

Venus appears so bright for two reasons. One, it comes closer to Earth than any other planet. And two, it is permanently covered in clouds, which reflect sunlight well.

Thick clouds permanently hide Venus's surface from view.

Like Mercury, Venus appears to change size over time. This happens as it makes its orbit around the Sun. It looks smallest when it lies on the other side of the Sun viewed from Earth, and looks largest when it comes between the Sun and Earth. The shape of Venus also appears to change as it circles in its orbit. These changing shapes are its phases (see page 170).

Again like Mercury, Venus's orbit sometimes takes it in a path so that it appears to travel across the surface of the Sun. This event is known as the transit of Venus. Such transits are very rare. The latest one took place in 2004.

At its closest, Venus is only about 26 million miles (42 million km) away from Earth. There are telescopes powerful enough to allow us to see features on Venus's surface. Even so, we are unable to because of the thick clouds that cover the planet.

Astronomers therefore use radar to study the surface. Radar uses beams of radio waves, which pass easily through clouds. Radar space probes, such as Magellan, beam down radio waves and pick up the reflections, or echoes. From the pattern of echoes, a detailed picture of the surface can be constructed.

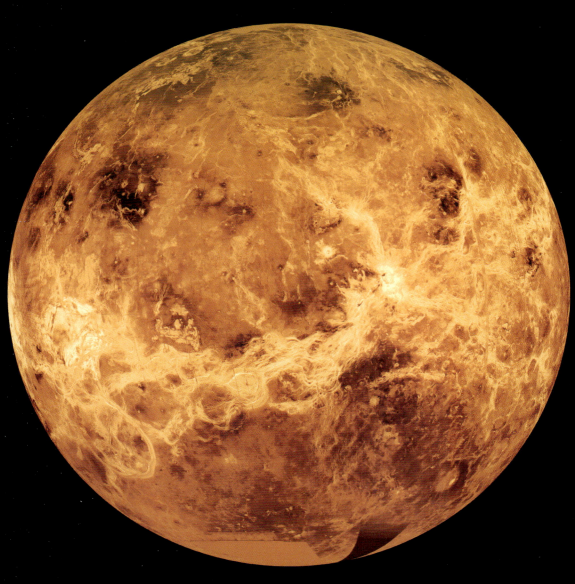

The Magellan probe used radar to peer through Venus's cloudy atmosphere. It spied a hot, dry, and barren landscape, dominated by volcanoes and the vast flows of lava that poured out of them.

Looking at Venus

Venus is sometimes called Earth's twin. Early in the 19th century, people thought that Venus might be a habitable world, similar to Earth but probably hotter. They imagined the planet covered with steamy swamps and populated by the kind of creatures that lived in the steamy swamps of Earth hundreds of millions of years ago.

But close examination by space probes have shown that Venus and Earth have little in common except their size. They have vastly different temperatures, atmospheres, and surfaces.

Slow Spinner

Venus orbits the Sun in almost a perfect circle, unlike Mercury, which follows an oval path around the Sun. It takes the planet 225 Earth-days to circle round the Sun once — this is Venus's "year."

Like all planets, Venus spins on its axis as it travels in its orbit around the Sun. But it does so very slowly — slower, in fact, than any other planet. It takes 243 Earth-days to spin around

VENUS DATA

Diameter: 7,521 miles (12,104 km)

Average distance from Sun: 67,000,000 miles (108,000,000 km)

Mass (Earth=1): 0.8

Density (water=1): 5.2

Spins on axis in: 243 days

Circles round Sun in: 225 days

Number of moons: 0

once. It spins around in the opposite direction to the other planets. They rotate from west to east, but Venus spins from east to west. If you lived on Venus, you would see the Sun rise in the west and set in the east — the exact opposite of what happens on Earth.

Below: The different phases of Venus. The size and shape of the planet we see changes as it moves closer or farther away from Earth. It is smallest when it is "full" (full circle).

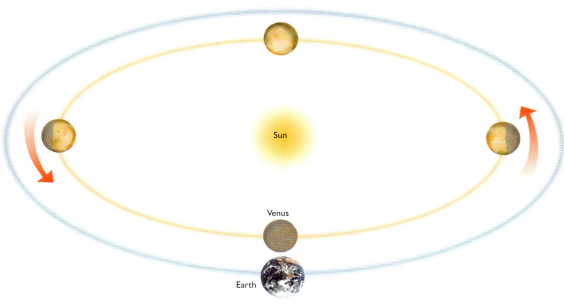

Sun

Venus

Earth

Crushing Pressure

However, if you did live on Venus, you would probably never be able to see the Sun rise and set. This is because the planet is always covered by thick white clouds that move rapidly through the atmosphere.

Venus has a very thick (dense) atmosphere — much thicker than Earth's atmosphere. The pressure (force) of the atmosphere at ground level on Venus is about 90 times greater than it is on Earth. Without some sort of suit or equipment that would make an adjustment to the pressure, a visitor to Venus would be crushed to death.

The main reason why Venus has such a thick and heavy atmosphere is because the atmosphere is made up almost entirely of carbon dioxide. Carbon dioxide is much heavier than the main two gases in Earth's atmosphere — nitrogen and oxygen.

Above: From Earth, we see different amounts of Venus lit up by the Sun as it travels in its orbit. In other words, it shows phases, like the moon does.

Venus's atmosphere also contains traces of water vapour (water in the form of a gas) and poisonous gases, such as sulphur dioxide and hydrogen chloride. All these gases, along with the carbon dioxide, have been spewed out by the many volcanoes that have erupted on Venus over the years.

Below: The most of Venus's very dense (thick) atmosphere consists of carbon dioxide.

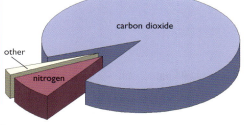

carbon dioxide

other

nitrogen

Top Temperature

Venus is about 25 million miles (40 million km) closer to the Sun than is Earth, which means it is extremely warm. Space probes have found that temperatures on the planet can reach as high as 480°C — hotter even than on Mercury. While Mercury is hot on one side and cold on the other, Venus is hot all over.

Like a Greenhouse

Venus stays so hot because its thick atmosphere traps or holds in heat, much like a garden greenhouse does. This is called the "greenhouse effect."

The planet is heated by sunlight, which passes through the atmosphere and heats the surface. The surface warms up and gives off heat rays. But the atmosphere prevents rays from escaping into space. They are absorbed by the carbon dioxide, water vapour, and other gases. Because the heat cannot escape, it builds up, making Venus the hottest planet in the solar system.

Acid Clouds

Venus would be even hotter if it were not for the haze and clouds in the atmosphere. On Earth, clouds are made up of tiny droplets of water or ice crystals. On Venus, however, they are made up of droplets of sulphuric acid. This is the kind of acid found in car batteries.

These acid clouds formed from the sulphur dioxide gas given off by the

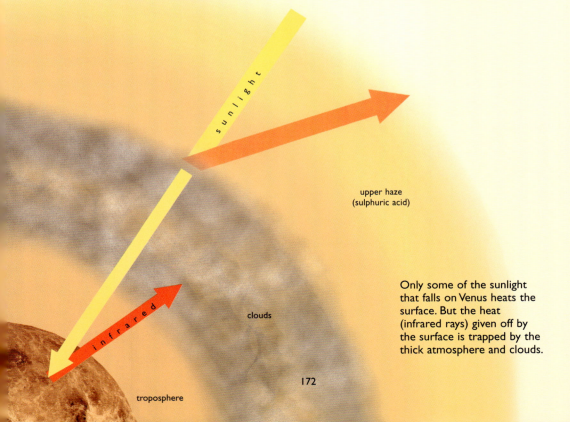

sunlight

upper haze
(sulphuric acid)

infrared

clouds

Only some of the sunlight that falls on Venus heats the surface. But the heat (infrared rays) given off by the surface is trapped by the thick atmosphere and clouds.

troposphere

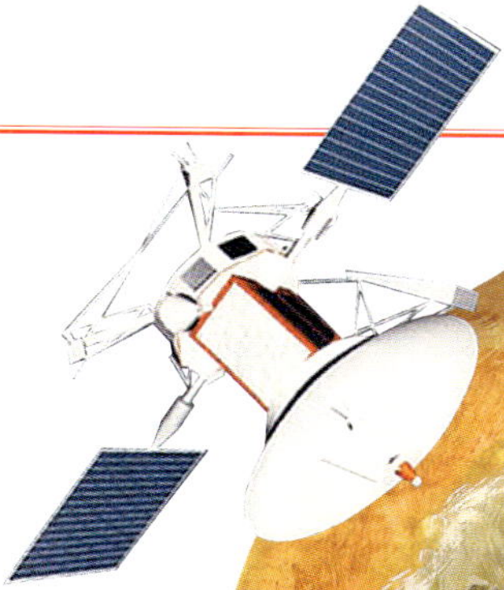

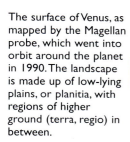

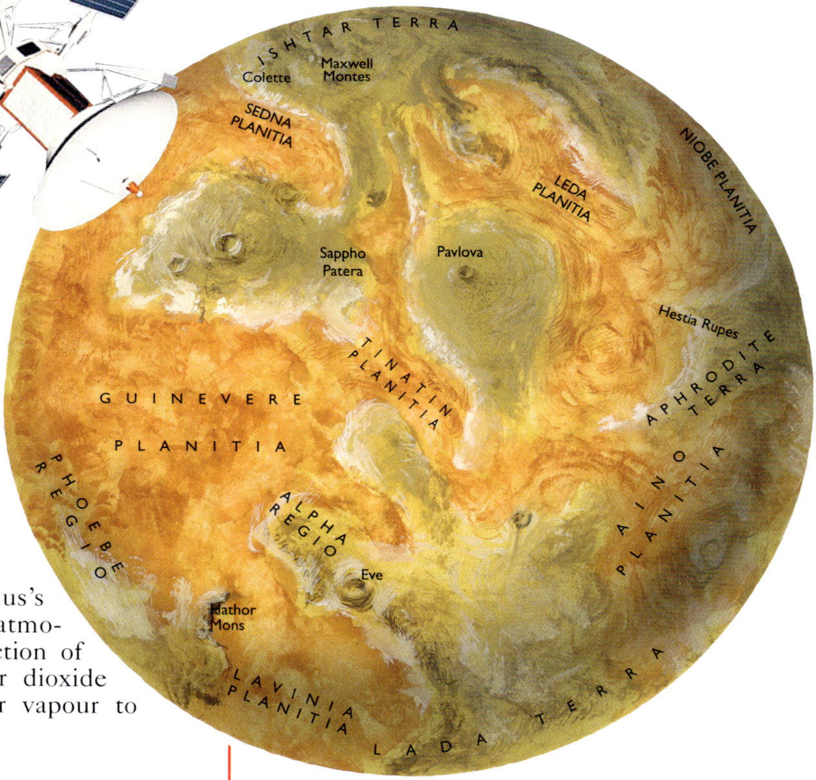

The surface of Venus, as mapped by the Magellan probe, which went into orbit around the planet in 1990. The landscape is made up of low-lying plains, or planitia, with regions of higher ground (terra, regio) in between.

Map labels: ISHTAR TERRA · Colette · Maxwell Montes · SEDNA PLANITIA · NIOBE PLANITIA · LEDA PLANITIA · Sappho Patera · Pavlova · Hestia Rupes · TINATIN PLANITIA · GUINEVERE PLANITIA · APHRODITE TERRA · PHOEBE REGIO · ALPHA REGIO · Eve · LAINO PLANITIA · Hathor Mons · LAVINIA PLANITIA · LADA TERRA

eruption of Venus's volcanoes. In the atmosphere, under the action of sunlight, the sulphur dioxide combines with water vapour to form the acid.

Venus Landscapes

Although clouds hide the surface of Venus from view through a telescope, radar scans from Earth and from spacecraft in orbit around Venus have told scientists much about Venus's surface.

Low, rolling plains cover about two-thirds of the surface of Venus. They have been likened to Earth's ocean basins. Venus has just two main upland areas, similar to Earth's continents.

The largest continent lies close to Venus's equator. It is named Aphrodite Terra — Aphrodite being the Greek name for the goddess of love and beauty, and Terra meaning "earth" or "land." This continent stretches for about 6,000 miles (9,500 km) and covers an area about the size of Africa. It is cleft by deep valleys, known as chasma. Deepest is Diana Chasma, which is more than 2.5 miles (4 km) deep and nearly 200 miles (300 km) wide in places.

The other continent lies further north. It is named Ishtar Terra, after the Babylonian goddess of love. It is about the same size as the United States. This continent is noted for its lofty mountains, including Maxwell Montes. They rise to heights of more than 5 miles (8 km) — about the same height as Earth's highest mountain, Mount Everest.

Venus's Volcanoes

Venus is covered with thousands of volcanoes. They have erupted all over the planet for millions of years, helping form its characteristic flat, rolling landscape. Many of them may still be active.

Most of the volcanoes found on Venus are the type known on Earth as shield volcanoes. The volcanoes on the Hawaiian islands, for example, are this type. Shield volcanoes are noted for their runny lava, which can travel a very long way. Venus's vast plains have been created by long and repeated lava flows.

Pancakes and Spiders

Volcanic activity on Venus has also produced features that can be found nowhere else in the solar system. They include flat circular domes, which look rather like pancakes. They are typically around 15 miles (25 km) across. Astronomers think that the pancakes formed when thick lava bubbled up through cracks in the surface. This lava did not flow far because it was so viscous (sticky).

There are other strange formations on Venus that look like spiders' webs. They are called "arachnoids," after a word that means "spider." Arachnoids have a central crater surrounded by a network of fine lines. These appear to be fractures (cracks) in the surface, made by molten rock pushing up from below.

Coronae (meaning "crowns") are other volcanic features unique to Venus. They are circular structures surrounded by many ridges and valleys and networks of fractures.

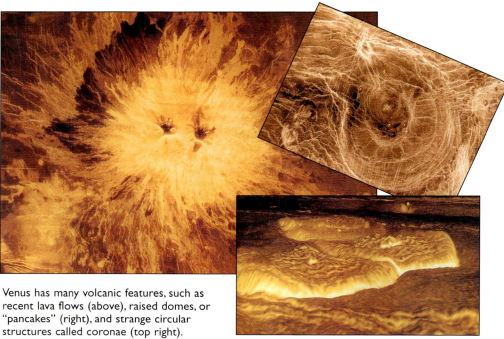

Venus has many volcanic features, such as recent lava flows (above), raised domes, or "pancakes" (right), and strange circular structures called coronae (top right).

Venus's Craters

Like all the inner planets, Venus was bombarded heavily by meteorites soon after it formed. They would have dug out craters, like those we see on Mercury today. But since then Venus's volcanoes have been at work, constantly pouring out lava that has filled in the ancient craters.

Today, the relatively few craters that can be seen on Venus are all quite young. They show up clearly against the older lava plains. There are some really spectacular craters, some more than 150 miles (250 km) across. In pictures, we see them surrounded by fresh material dug out when meteorites smashed into the surface.

The atmosphere helps protect Venus from smaller meteorites. They burn up like meteors or break apart when they hit the atmosphere. Sometimes we see craters with several pits inside them, showing where separate pieces of meteorite have landed. This kind of crater is not found on Mercury or the Moon because they have no atmosphere to break up the incoming missiles.

Below: Volcanoes by the hundreds dot Venus's surface. Many are probably still active.

Bottom: Only a few impact craters are found on Venus. This one measures about 30 miles (50 km) across.

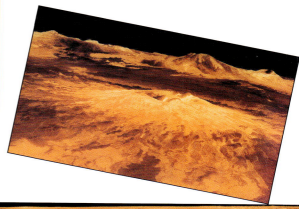

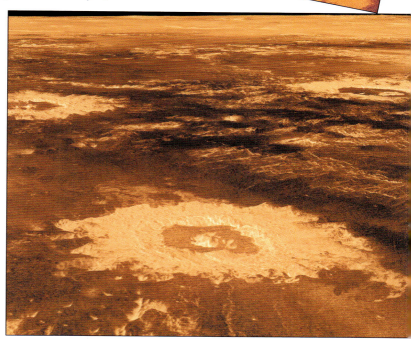

PLANET EARTH

Our home planet, Earth, is the largest of the inner planets, but it is a dwarf compared with the giant outer planets, such as Jupiter. It behaves like a typical planet, spinning around on its axis as it orbits the Sun. What makes Earth unique among planets is that it teems with life.

Earth provides a comfortable home for millions of different species (kinds) of living things, from tiny microscopic plants to gigantic animals, such as whales. Whales are the biggest animals that have ever lived, bigger even than the dinosaurs. They are masters of their environment — the oceans — where life is believed to have started billions of years ago.

Why is there so much life on Earth but none on the other planets? The answer seems to be that Earth just happens to be located in the right place in the solar system and just happens to be the right size.

Earth is located in the solar system in what is sometimes called the life zone. It receives just the right amounts of heat and light from the Sun to provide just the right conditions for life to flourish. In particular, its location allows water to exist as a liquid, and water is needed by almost all living things.

Earth has the right size and make-up to take advantage of conditions in the life zone. They give it a large enough mass and gravitational pull to hold onto a layer of gases, or atmosphere. The atmosphere provides oxygen for living things to breathe. It also acts like a blanket to help keep Earth warm, and to shield it against deadly radiation from space.

Earth is often called a living planet for another reason. Like a living thing, it is constantly changing. On its outside, the weather, running water, and other forces are continually wearing away and reshaping Earth's surface. Beneath the surface, on the inside, movements in the rocks cause the continents to move, volcanoes to erupt, and earthquakes to take place.

Left: This is a volcanic rock called pumice, which is widely found on Earth. It formed from lava that poured out of ancient volcanoes.

A view of our very watery planet Earth, showing the vast Pacific Ocean and the continents of North and South America.

Looking at Earth

Earth is the third planet out from the Sun, located between Venus and Mars. Like these and all the other planets, Earth has two motions in space. It both spins around — rotates — on its axis and it circles around the Sun.

We do not feel the Earth move, of course, because we are moving with it. But it moves very, very fast. A person standing on the Equator, is travelling around the centre of the Earth at a speed of 1,000 mph (1,600 km/h). At the same time, the inhabitants of Earth are hurtling through space in orbit around the Sun at a speed of more than 66,000 mph (105,000 km/h).

Marking Time

Because Earth rotates, the Sun and the other stars seem to circle around Earth. This is why our early ancestors thought that Earth was the centre of the universe.

Every day at noon, the Sun reaches its highest point in the sky. The time between two noons is always 24 hours, or 1 day. The day is one of our basic units of time.

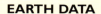

EARTH DATA

Diameter: 7,927 miles (12,756 km)

Average distance from Sun: 93,000,000 miles (150,000,000 km)

Mass (Earth=1): 1

Density (water=1): 5.5

Spins on axis in: 23.94 hours

Circles round Sun in: 365.25 days

Number of moons: 1

Just as Earth always spins completely around in the same period of time, so Earth always takes the same time to complete its orbit around the Sun — a little over 365 days. This period is our other basic unit of time, the year.

Earth spins around on its axis at an angle of 23½°.

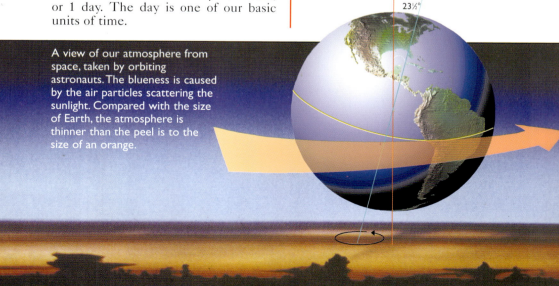

A view of our atmosphere from space, taken by orbiting astronauts. The blueness is caused by the air particles scattering the sunlight. Compared with the size of Earth, the atmosphere is thinner than the peel is to the size of an orange.

23½°

The Tilting Earth

Earth does not spin in an upright position as it travels in its orbit around the Sun. Its axis is slightly tilted (by 23½ degrees). Earth always stays tilted in the same direction in space. This means that places on Earth are tilted more toward the Sun — and are hotter — at some times of the year than at others.

As a result, most places experience regular changes in temperature and the weather throughout the course of the year, bringing about what we call the seasons. In many parts of the world, there are four seasons — spring, summer, fall (autumn), and winter. Mars also has seasons.

The Atmosphere

By weather, we mean the state of the atmosphere — whether it is hot or cold, dry or wet, still or windy, clear or cloudy. The atmosphere is the layer of air that surrounds Earth. It is thickest (densest) near the ground in a layer called the troposphere by meteorologists, the scientists who study the weather.

Above the troposphere are other layers, called the stratosphere, mesosphere, thermosphere, and exosphere. Going up through these layers, the air gets thinner and thinner until it merges into space at a height of about 300 miles (500 km).

Right: Earth's atmosphere extends up for more than 300 miles (500 km), where it merges into space. Some spacecraft orbit at the edge of the atmosphere. Glowing lights called the aurora occur high up too. Meteors burn up lower down.

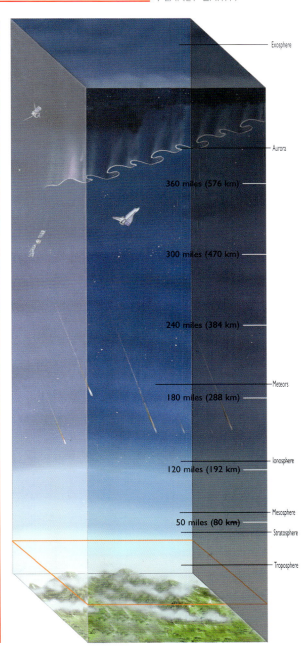

Exosphere

Aurora

360 miles (576 km)

300 miles (470 km)

240 miles (384 km)

Meteors

180 miles (288 km)

Ionosphere

120 miles (192 km)

Mesosphere

50 miles (80 km)

Stratosphere

Troposphere

179

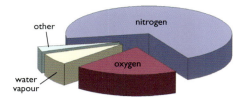

Earth's atmosphere is made up mostly of nitrogen and oxygen.

Gases in the Atmosphere

The two main gases in Earth's atmosphere are nitrogen (about 78 percent) and oxygen (about 21 percent). Oxygen is by far the most important of the two, because it is the gas humans and almost all living things must breathe to stay alive.

There are traces of many other gases in the atmosphere. One is carbon dioxide, the gas that humans breathe out. It plays an important part in life on Earth because plants need to take it in to make food. In sunlight, they combine carbon dioxide with water to make sugar, their food. This process is called photosynthesis, which means literally "to make light."

Another important trace gas in the atmosphere is water vapour — water in the form of a gas. Water plays a vital part in our weather. Water circulates constantly between Earth's surface and atmosphere. This is called the water cycle.

Most water gets into the air through evaporation, which is when the heat of the Sun turns surface water, such as in oceans or lakes, to vapour. In the air, the vapour cools and turns into droplets of water or ice crystals, creating mists, clouds, rain, and snow.

Water is found nearly everywhere on Earth — in the ground, on the surface, and in the atmosphere. In the atmosphere, water is found as vapour (gas) and as tiny droplets in clouds. When the droplets get big enough, they fall to the ground as rain. Vapour, clouds, and rain are features of Earth's never-ending water cycle.

Land and Sea

Earth has quite a different surface from all the other planets. Large areas of its rocky crust (outer layer) are hidden under water, which forms vast oceans. The oceans cover more than 70 percent of Earth's surface, more than twice the area of the dry land. The Pacific Ocean alone covers an area equal to that of all the land masses put together.

The main land masses are the seven continents: North America, South America, Europe, Asia, Africa, Australia, and Antarctica. Earth's crust is thickest under the continents, averaging about 25 miles (40 km) in thickness. The crust under the oceans is much thinner — only about 6 miles (10 km).

Drifting Continents

The map shows the location and shapes of the continents as we know them today. But they have not always looked like this. About 200 million years ago, all these land masses were joined together to form one supercontinent, which geologists call Pangaea.

Since then, they have split apart and drifted across the face of the Earth to where they are located today. They will continue to drift in the future, so that in many millions of years' time, the arrangement of Earth's land masses will be quite different from what it is today.

How does this continental drift occur? It happens because Earth's crust is not a single layer of rock but is made up of many sections, called plates. Each plate is able to move. The continents sit on plates and move with them. This theory of moving plates is known as plate tectonics.

170 million years ago

50 million years ago

Right: Earth's surface as it is today, made up of continents and islands set in a vast saltwater ocean. But the Earth has not always looked like this. It has changed drastically over the last 150 million years, and is changing still.

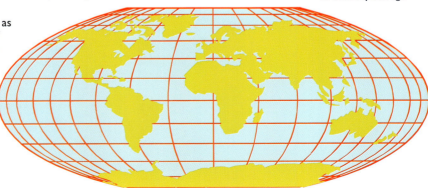

Wandering Plates

The rocks that form the plates of Earth's crust are hard and rigid. But beneath the crust, the rocks are much hotter and softer and can flow slowly, something like chewing gum or putty. Heat from inside the Earth sets up currents in these rocks, and it is these that carry the plates along.

The plates fit together something like a jigsaw puzzle. Where they meet can be the site of all kinds of unusual geological events. Two great plates meet in the middle of the South Atlantic Ocean, for example. They are pulling apart, causing the ocean to grow wider. This also happens in Earth's other oceans and is called sea-floor spreading.

In other parts of the world, plates are colliding. This is happening, for example, off the west coast of South America. An ocean plate is pushing against the plate carrying South America. Something has to give, and land at the edge of South America rides up. This is what created the great mountain range known as the Andes.

Volcanoes and Earthquakes

At the same time, the ocean plate dips down under the land plate. As it does so, it rubs against the surrounding rocks. This can make the ground shake — an earthquake. The friction also makes the plate material heat up and melt.

Pressure forces the molten rock upward through cracks to the surface. This is what volcanoes are and explains why there are so many volcanoes in the Andes.

Volcanoes or earthquakes or both occur at most of the other places in the world where plates meet. The most active regions for volcanoes are around the outer rim of the Pacific Ocean, where plates collide head-on. Some 300 volcanoes are found in the so-called Ring of Fire.

Slow-moving currents in the rocks deep underground carry along plates of Earth's crust. Under the oceans, plates are moving apart, causing them to widen. Where ocean plates come up against plates carrying the continents, the land rides up, creating mountains.

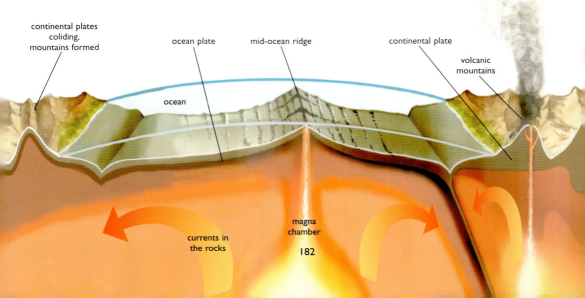

continental plates coliding, mountains formed

ocean plate

mid-ocean ridge

continental plate

volcanic mountains

ocean

magna chamber

currents in the rocks

Under the Weather

Volcanoes and earthquakes are not the only things that help reshape the Earth's surface. The weather and the flow of water contribute as well. They cause the surface gradually to erode, or wear away. Flowing water is one of the primary agents of erosion. Rivers cut deep into rock to create such natural wonders as the Grand Canyon.

Water carries away the material it gouges out of rock and deposits it elsewhere as sediment. Over millions of years, sediment can build up and turn into rock. Most of the rocks that cover our Earth now are sedimentary rocks, which formed from deposits laid down in ancient seas.

On the other planets, all the rocks are volcanic, meaning that they formed from molten rock. However, Mars may be an exception. Recent evidence shows that water did once flow on the planet, so there may well be sedimentary rocks there, too.

Above: An Hawaiian volcano erupts in a spectacular fireworks display.

Below: The Grand Canyon has been worn away by the Colourado River over millions of years.

THE RED PLANET

Mars shines in the night sky with a fiery reddish-orange colour. The colour reminded the ancient Romans of fire and blood, so they named Mars after their god of war. People once thought that there might be life on Mars — even intelligent life. But space probes have found no signs of it.

Mars comes nearer to Earth than any other planet except Venus. It is most visible in the night sky about every 26 months. That is when it comes closest to Earth as the two planets orbit the Sun. Astronomers call these times oppositions.

Every 15 years or so, Mars gets especially close to Earth, coming within about 35 million miles (56 million km). Then it becomes one of the brightest objects in the night sky rivaling even Jupiter in brightness. It is easy to tell which planet is which, however, because Mars is reddish and Jupiter is pure white.

Through a telescope, astronomers can see some details on Mars's surface. There are dark markings, which seem to change shape over time. And there are white regions at the north and south poles. They look like the areas of ice found at Earth's North and South Poles.

Mars is also similar to Earth in other ways. It completes its rotation on its axis in a little more than 24½ hours. In other words, its "day" is only about ½ hour longer than our own.

Also, Mars has seasons like Earth does. As on Earth, the seasons are caused by the tilt of the planet's axis in space. This tilt means that parts of Mars are closer to the Sun at some times than at others. Those that are tilted most toward the Sun enjoy their warmest season — summer. When they are tilted farthest away from the Sun, they experience their coldest season — winter.

On Mars, the seasons are nearly twice as long as they are on Earth because Mars's "year" (the time it takes to circle the Sun) is nearly twice as long as our year. Because Mars is so much farther from the Sun than Earth is, temperatures during their seasons are much colder than they are on Earth.

Mars boasts massive volcanoes bigger than Mount Everest. Ancient lava flows cover much of the northern hemisphere (half) of Mars.

The red planet Mars, looks reddish
in the night sky and also up close.
Like Earth, it has ice caps and
clouds in its thin atmosphere.

Mars's Atmosphere

Mars is similar to Earth in one other way — it has an atmosphere. This can be told from Earth because at times astronomers cannot see any markings on Mars through their telescopes. The markings have been blotted out by huge dust storms that sweep over the whole planet. Without an atmosphere, there would not be any dust storms.

Mars's atmosphere, however, is nothing like Earth's. There is very little of it — scientists say that it is very thin. It presses down on the surface with only one percent of the pressure (force) of Earth's atmosphere.

Even though its atmosphere is very thin, strong winds sometimes blow on Mars. This is what whips up dust from the surface and creates planet-wide dust storms. On occasions, winds may blow at speeds of 200 mph (300 km/h) or more — these are the speeds winds reach in tornadoes on Earth.

Gases in the Atmosphere

The atmosphere on Mars also has a much different chemical make-up than Earth's. It is made up mainly of carbon dioxide, which is the gas given off by the burning of fuels such as oil. The carbon dioxide on Mars, of course, did not come from burning fuels — it was given off when volcanoes erupted on the planet.

Mars's atmosphere consists mainly of carbon dioxide.

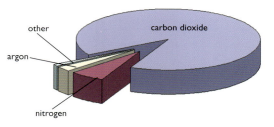

other
argon
carbon dioxide
nitrogen

MARS DATA

Diameter: 4,221 miles (6,792 km)
Average distance from Sun:
142,000,000 miles (228,000,000 km)
Mass (Earth=1): 0.1
Density (water=1): 3.9
Spins on axis in: 24.63 hours
Circles round Sun in: 687 days
Number of moons: 2

There are also tiny amounts of other gases in Mars's atmosphere, including nitrogen and oxygen. These are the two gases that make up our own atmosphere. But there is nowhere near enough oxygen for humans to breathe on Mars. An astronaut who explored Mars would have to wear a spacesuit with breathing apparatus that provided oxygen to breathe.

Cloudy Skies

Traces of yet another gas can be found in Mars's atmosphere — water vapour. This is water in the form of a gas. On Earth, the heat of the Sun turns water into vapour, which rises into the air. This process is called evaporation. There is no liquid water on Mars, but there is frozen water, or ice. The heat of the Sun turns this into vapour, which rises into the atmosphere.

On Earth, as water vapour rises, it gets colder, and it turns back into little droplets of water. These droplets group together to form white clouds. A similar thing happens on Mars, only the clouds that form are made up of tiny ice crystals, not water droplets, because the temperature is so low. Clouds can often be seen on Mars clinging to the sides of the tall volcanoes. And low clouds, or mists, often form in Mars's many valleys.

Above: Part of Mars's southern hemisphere. The circular white patch is a vast region covered in a water-ice mist.

Above: Dust storms often blow up in the sandy desert regions of Mars (lower part of picture).

Sunset on Mars, snapped by one of the Viking landers, which landed on the Red Planet in 1976.

Above: The northern ice cap of Mars shows up in this picture taken by the Hubble Space Telescope.

Below: Much of Mars consists of sandy plains, strewn with rocks. This Viking picture shows Chryse, the Plain of Gold.

Temperatures on Mars

Mars is on average about 50 million miles (80 million km) further away from the Sun than Earth. This means that it is much colder than Earth.

On any planet, the hottest region is near its "waist," or equator. This is the part that receives the most heat from the Sun. Around Earth's equator, temperatures regularly rise to more than 104°F (40°C). But on Mars, temperatures at the equator struggle to rise above freezing point 32°F (0°C). But they may rise briefly to about room temperature 70°F (21°C) in mid-summer.

Elsewhere on the planet, temperatures are low all the time. When the Viking 1 probe landed on the Martian plain Chryse in 1976, it found that the temperature around midday was only about −20°F (−30°C). During the night, the temperature fell to below −110°F (−80°C). On Mars, the temperature falls rapidly after the Sun sets because the atmosphere is so thin. Earth's thick

atmosphere acts like a blanket to help hold in the warmth of the day. The coolest parts of Mars are the regions around the north and south poles. They are at their coldest in winter, when the tilt of Mars leaves them farthest away from the Sun. Then, temperatures drop to −240°F (−150°C) or below.

The Ice Caps

The main features of the polar regions of Mars are the ice caps, which can be seen through telescopes from Earth. They change size with the Martian seasons.

As to be expected, the ice caps are largest in winter, when the weather is intensely cold. As spring comes and the weather warms up, they start to melt and shrink in size. The caps are smallest in summer. Then, as autumn comes and temperatures begin to fall, ice freezes out of the atmosphere, and the caps start to grow again.

There is always some ice left at the poles even in midsummer. Astronomers believe that this is mainly water ice — frozen water. The icy material that makes the ice caps grow as winter

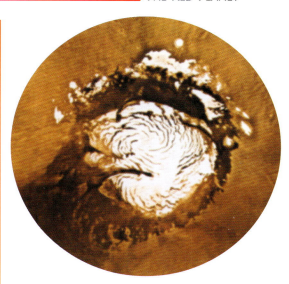

A close-up of Mars's north polar ice cap. In summer, the cap measures up to 370 miles (600 km) across and is mainly water ice. In winter, the cap expands as frozen carbon dioxide settles over it.

comes is probably frozen carbon dioxide, or "dry ice."

On the Plains of Gold

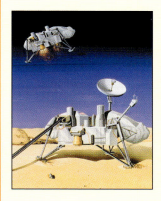

On June 19, 1976, the first of two identical Viking probes slipped into orbit around Mars. It had set off from Earth just 10 months before. Viking 1 then spent a month looking for a suitable landing site, and on July 20 it released a lander. Using the drag of the planet's atmosphere, rockets, and parachutes, the lander made a soft touchdown on a region of Mars known as Chryse Planitia — the Plains of Gold. Within seconds, its cameras began photographing the surrounding area. It looked like a stony desert on Earth, with many small rocks scattered around a sandy landscape. This was the first clear close-up view of another planet sent back to Earth. Soon the lander extended a meteorology (weather) boom and began sending back measurements of wind speed, atmospheric pressure, and temperature. The lander became Mars's first weather station.

Left: The two Viking craft (inset) that went into orbit around Mars in 1976 mapped the whole planet in considerable detail, much of it in colour.

Opposite top: Much of Mars's southern hemisphere is covered with craters, large and small.

Looking at the Landscape

The surface of Mars boasts all kinds of fascinating features — vast deserts, gigantic volcanoes, long valleys, huge basins, and craters large and small.

Like all the inner planets, Mars was bombarded by huge lumps of rock — meteorites — early in its history. This bombardment scarred the planet with pits, or craters. Much of the southern hemisphere (half) of the planet is still dotted with ancient craters. They make this region of Mars look rather like the highland regions of the Moon.

The craters on Mars look much like those on the Moon. They have raised walls and low floors. The larger ones have a small mountain range in the centre. In some, the impact (blow) that formed the crater melted the rocks. Molten material flowed into the surrounding area, forming a noticeable ring around the crater. This kind of thing never happened on the Moon.

The Basins

Two very large meteorites or asteroids hit the southern hemisphere of Mars long ago, gouging out two huge basins. They are both surrounded by low mountain ranges, thrown up by the force of the impacts.

The largest basin, named Hellas, is more than 1,350 miles (2,200 km) across. The floor of the basin is flat and resembles a sandy desert on Earth. The other basin, Argyre, is only about half the size of Hellas. Argyre becomes very conspicuous at times when it fills with frost.

The floor of the gigantic Hellas basin. It is the lowest region on Mars, about 2 miles (3 km) lower than the surrounding surface.

The Volcanoes

The heavily cratered region of the southern hemisphere of Mars extends some way into the northern hemisphere. But vast plains cover most of the north. They were created by the lava flows from innumerable volcanoes.

The greatest of these volcanoes are found just north of Mars's equator, on a huge bulge on the surface in a region astronomers call Tharsis. There are four major volcanoes. The biggest of them all is Olympus Mons (Mount Olympus).

The biggest volcano we know in the solar system, Mars's Olympus Mons. Its central crater is over 40 miles (70 km) across.

This gigantic mountain soars to a height of around 17 miles (27 km). This is four times the height of the highest peak in North America, Denali (Mt. McKinley) in Alaska. Olympus Mons is very broad at the base, measuring more than 400 miles (640 km) across. It is a type of volcano called a shield volcano, like those on Hawaii such as Mauna Loa.

191

Mars's Grand Canyon, Mariner Valley. Close-up pictures of the valley show landslides, signs of erosion by the wind, and what seem to be water channels.

Canyonlands

Just to the east of the Tharsis Bulge is the biggest natural feature on Mars, a huge valley that runs close to the equator for more than 2,500 miles (4,000 km). It is called Mariner Valley (Valles Marineris), after the Mariner space probe that discovered it.

The main part of the valley is up to 250 miles (400 km) wide and 6 miles (10 km) deep. It has many side branches that end in steep-sided canyons. It has often been called Mars's Grand Canyon, after the spectacular natural feature of that name in Arizona. But it is four times deeper, six times wider, and ten times longer than the Grand Canyon on Earth.

The only comparable valley on Earth is the Great Rift Valley in East Africa, which formed when two blocks of the Earth's surface moved apart. Mariner Valley probably formed in a similar way, when the Martian surface cracked. Maybe it split apart when the Tharsis Bulge forced its way up.

Mariner Valley certainly did not form as the Grand Canyon did. The Grand Canyon was formed over many millions of years as a result of the Colourado River cutting through the surface rocks. Mariner Valley does, however, show signs that water has flowed through the canyons and gorges at some time in the past.

Sandblasting

The surface of Mars has been shaped mainly by volcanoes, internal forces, and meteorites. But other forces have also been at work — in particular, the wind. In Mars's desert regions, the wind creates rippling dunes — waves in the sandy soil.

The strong winds on Mars also whip

up sand and dust from the surface. They "sandblast" the mountains, craters, and other landforms, slowly wearing them down. Such wearing away, or erosion, is common in desert regions on Earth.

Most erosion on Earth is caused by flowing water, however. There are signs that flowing water has also caused some erosion on Mars.

The Channels

Mariner Valley and the large canyons on Mars were formed by massive movements in the planet's crust. But most of the smaller valleys and channels that snake over the surface of Mars were obviously formed in a different way. Some were made by lava flowing from ancient volcanoes. Similar channels are found on the Moon — a classic example is Alpine Valley, which cuts through the lunar Alps.

But many channels look as though they have been made by flowing water.

A region of Mars known as Mangala Vallis, showing the boundary between higher, older crust and a smooth plain. The deep channels seen here were probably cut by flowing water.

Some look like dried-up riverbeds. They display the typical features of river formation — starting out narrow (where the river rises), then gradually broadening out and splitting into branches (tributaries) downstream. Other channels start broad, as if created in massive floods. Elsewhere, we find teardrop-shaped patterns around craters, which look as if they have been made by rivers flowing round them.

Water Everywhere

Astronomers are now almost certain that the most of the channels they have spotted on Mars were made by flowing water long ago. They think that the vast desert plains of northern Mars once might even have been oceans.

This supports the idea that Mars once had a warmer climate and a thicker atmosphere. This would have been long ago, when volcanic eruptions were taking place all over the planet. But over time, the climate cooled and the atmosphere slowly escaped into space, taking with it most of the planet's moisture.

The moisture that remains on Mars today is in the form of ice at the poles and clouds, mists, and frost elsewhere. Many astronomers now believe that there is a lot of moisture on Mars that has not yet been detected. Most likely, it is trapped as ice beneath the surface. A similar thing happens in Earth tundra regions, which are the cold land areas near the North and South poles. There, ground below the surface stays permanently frozen — it is known as permafrost.

Permafrost may well be found all over Mars. If so, it will provide a much needed supply of water when the human exploration of the Red Planet begins later in the 21st century.

Saturn and it's moons, a montage of
images prepared from Voyager 1.
Clockwise, the moons are as follows:
Dione (in front of Saturn), Enceladus,
Rhea, Titan, Mimas, and Tethys.

Exploring the Far Planets

Earth is one of nine bodies that are called planets, which circle, or orbit, the Sun. With thousands of other bodies, the planets make up what is known as the solar system. These other bodies include large lumps of rock and metal called asteroids and icy bodies called comets, which occasionally blaze across our skies. In distance from the Sun, with the nearest listed first, the planets are Mercury, Venus, Earth, Mars, Jupiter, Saturn, Uranus, Neptune, and Pluto. Earth and its three closest neighbours among the planets — Mercury, Venus, and Mars — lie relatively close together in the heart of the solar system. These planets are separated from each other by several tens of millions of miles, which in space is considered a relatively short distance. Together, they are often referred to as the near or inner planets.

Hundreds of millions of miles separate them and Jupiter, the next planet out. The planets beyond Jupiter are even farther away. Together, Jupiter, Saturn, Uranus, Neptune, and Pluto are referred to as the far or outer planets.

The outer planets differ greatly from the inner planets. The inner planets are small, and most of the outer planets are giant. The inner planets are rocky, and most of the outer planets are made up of gas and liquid. The inner planets have only three moons between them, but the outer planets have more than 60. Most of the outer planets have rings around them, and the inner planets do not.

Although Jupiter is very far from Earth, it is so big that we can see it shining brightly in the sky for many months of the year. When its orbit brings it close to Earth, Saturn also shines brightly in the sky, in part because of its magnificent rings. Both planets were known to ancient stargazers. But the three most distant planets were not. They are too far away to be clearly visible to the naked eye. Uranus was not discovered until 1781, Neptune not until 1846, and Pluto not until 1930.

Paths of the Planets

The five outer planets — Jupiter, Saturn, Uranus, Neptune, and Pluto — occupy a vast region of space. They are far apart. Saturn never comes closer to Jupiter than about 400 million miles (640 million km). Neptune never comes closer to Uranus than about 1 billion miles (1.6 billion km).

The farther out the planets are, the longer is their "year" — the time it takes to complete one orbit around the Sun. Their "years" vary from just under 12 Earth-years for Jupiter to more than 248 Earth-years for Pluto.

Oval Orbits

All the planets orbit the Sun not in circles but in paths that are oval, or elliptical, in shape. The orbits of all the outer planets except Pluto are only slightly oval — they are nearly circular. But Pluto's orbit is very oval. This means that its distance from the Sun varies widely during the course of its year. Sometimes it gets within 2.8 billion miles (4.5 billion km) of the Sun. At other times it wanders more than 4.5 billion miles (7.4 billion km) away.

Pluto's orbit is unusual in another way. All the other outer planets circle the Sun in much the same plane (flat sheet) in space. But Pluto travels quite a long way above and below this plane.

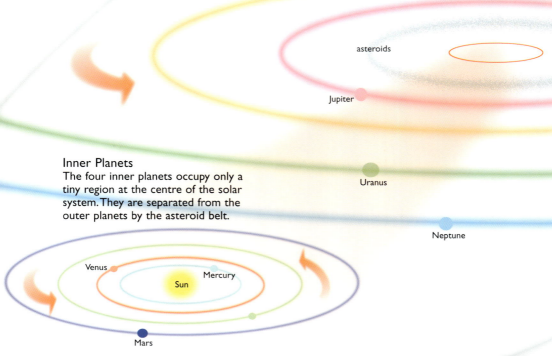

Outer Planets
The five outer planets are widely scattered in space. Like the inner planets, they all orbit the Sun in the same direction.

asteroids

Jupiter

Uranus

Neptune

Inner Planets
The four inner planets occupy only a tiny region at the centre of the solar system. They are separated from the outer planets by the asteroid belt.

Venus

Mercury

Sun

Mars

Planet Data

Planet	Av. distance from Sun million miles (km)	Diameter at equator in miles (km)	Completes Orbit (Length of Year)	Rotates around in (Length of Day)	Mass (Earth=1)(water=1)	Density	Number of moons
Jupiter	483 (778)	88,850 (142,980)	11.9 years	9.39 hours	318	1.3	16
Saturn	888 (1,429)	74,900 (120,540)	29.4 years	10.66 hours	95	0.7	18
Uranus	1,787 (2,875)	31,765 (51,120)	83.7 years	17.24 hours	15	1.3	18
Neptune	2,799 (4,504)	30,779 (49,530)	163.7 years	16.11 hours	17	1.6	8
Sun	—	865,000 (1,392,000)	—	25 days	330,000	1.4	—
Earth	93 (150)	7,927 (12,756)	1 year	23.93 hours	1	5.5	1

Pluto

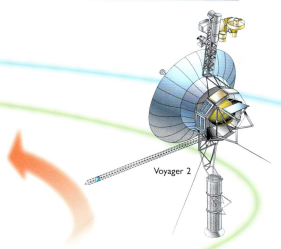

Voyager 2

Saturn

Exploring the Outer Planets

Considering how far away the outer planets are, we know an amazing amount about them. For example, we know that Jupiter's moon Io has active volcanoes. Saturn's rings are made up of millions of ringlets. Neptune has white-flecked clouds scurrying through its deep blue atmosphere. We have learned these things — and many others — from unmanned space probes that have explored the planets close-up.

Pioneer 10 was the first probe to travel to Jupiter. Pioneer 11 visited Saturn as well. But most of our knowledge of the outer planets has come from the Voyager 1 and 2 probes. Both were launched in 1977 and both visited Jupiter (1979) and Saturn (Voyager 1 in 1980, Voyager 2 in 1981). Then Voyager 2 went on to fly by Uranus (1986) and Neptune (1989).

Size and Structure

The outer planets are very different in size and make-up from the inner planets. Four of them — Jupiter, Saturn, Uranus, and Neptune — are often called gas giants because of their huge size and because they are made up mainly of gas. They have no solid surface. Pluto is different — it is a tiny ball of rock and ice.

Jupiter and Saturn are by far the biggest planets, and they have a similar make-up. They have a relatively thin layer of atmosphere, made up mainly of the gases hydrogen and helium. At the bottom of the atmosphere, high pressure forces the hydrogen gas to a liquid. Deeper down, even higher pressure turns the liquid hydrogen into a kind of liquid metal. This liquid metallic hydrogen reaches right down to the centre, or core, of the planets, which is probably made up of dense rock.

The other gas giants, Uranus and Neptune, are similar both in size and in make up. They have a deep atmosphere made up of hydrogen, helium, and methane gases. Underneath, there is a vast ocean made up of hot liquid containing water, methane, and ammonia. At the centre, they have a small rocky core.

How the Gas Giants Formed

The solar system came into being about 4.6 billion years ago. That was when the Sun was born and the planets formed around it — both the small rocky inner ones and the giant, gassy outer ones. So how is it that the inner and outer planets came to be so different in structure? Astronomers think that this happened soon after the Sun became a star.

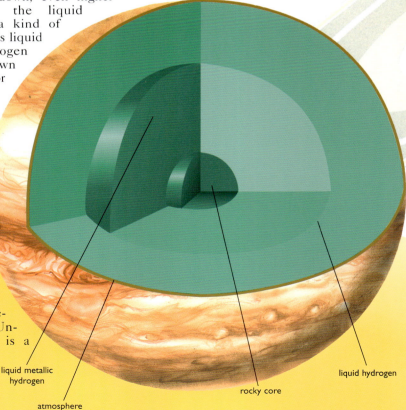

liquid metallic hydrogen

atmosphere

rocky core

liquid hydrogen

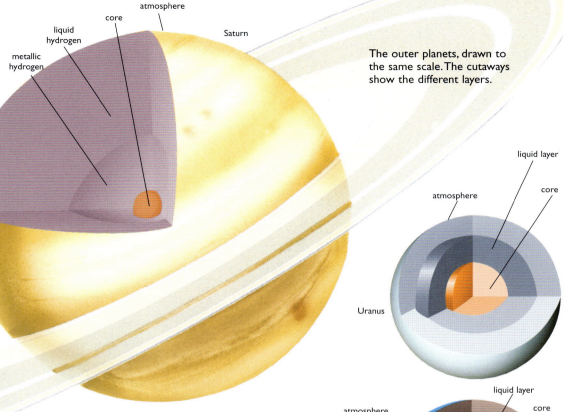

atmosphere

core

liquid
hydrogen

metallic
hydrogen

Saturn

The outer planets, drawn to
the same scale. The cutaways
show the different layers.

liquid layer

core

atmosphere

Uranus

liquid layer

core

atmosphere

Neptune

frozen methane,
nitrogen, and carbon
monoxide

Pluto

Pluto is so tiny that we
have had to magnify it
here about 25 times to
show its makeup.

core

frozen
water ice

In the process, the Sun blasted off its outer layers. The blast of matter blew like a mighty wind through the young solar system. It stripped away the gassy atmospheres from around the nearby inner planets, leaving them bare. The outer planets were not so affected in the same way because they were much farther away from the Sun. They kept their original atmospheres, and their gravity, or pull, and gradually attracted more and more of the gas that had been blasted out of the inner solar system. Over time, they grew into the giant bodies we find today.

199

Sun

GIANT JUPITER

Jupiter is by far the largest of the planets. A gigantic ball of gas and liquid, Jupiter is more than 11 times greater in diameter than Earth. It could fit within itself 1,300 bodies the size of Earth. Jupiter is named after the king of the gods in Roman mythology, and in terms of size it is the "king" of the planets.

Jupiter is the fifth planet in the solar system in distance from the Sun. It circles around the Sun at an average distance of about 500 million miles (800 million km) and takes nearly 12 years to circle the Sun once. Jupiter lies between the nearer planet Mars and the more distant Saturn, but the closest bodies to it are the smaller bodies known as asteroids. They circle the Sun in a broad ring, or belt, roughly halfway between the orbits of Mars and Jupiter.

Even though Jupiter never comes closer to Earth than about 400 million miles (640 million km), it shines brightly in the night sky. To people on Earth it appears as a brilliant white "star" during many months of the year. The brightest object in the night sky after the Moon and Venus, it far outshines the "real" stars. Mars can sometimes rival Jupiter in brightness, but it can easily be distinguished from Jupiter by its fiery orange or reddish colour.

Jupiter is a good object for observation. Using ordinary binoculars, you can see it as a circle, or disk, and you may be able to spot four of its moons as points of light on either side. This reminds us that Jupiter is the centre of its own miniature system of heavenly bodies, with at least 16 moons circling around it.

Through a telescope, Jupiter looks magnificent. Its disk is crossed with coloured bands and dotted with spots. When you watch the planet for some time, you can see the spots and other features move across the disk. This shows that Jupiter is spinning around very rapidly. Observations show that it spins, or rotates, completely on its axis in less than 10 hours — faster than any other planet. The period of time that it takes Jupiter (or any other planet) to complete a single rotation is the length of its day.

Left: Jupiter's moon Io is more brilliantly coloured than any other moon in the solar system.

White and reddish-orange bands of clouds fill Jupiter's thick atmosphere. Between these bands are great turbulent regions where currents in the atmosphere eddy this way and that. It is in these regions that violent storms take place.

The Belts and Zones

Space probes such as Pioneer 10 and 11, the two Voyagers, and Galileo have studied Jupiter in detail. They have shown that the coloured bands on Jupiter's disk are fast-moving clouds high in the planet's atmosphere.

The clouds have been drawn into parallel bands because the atmosphere moves so quickly. Astronomers call the dark bands of clouds "belts" and the light ones "zones."

The Cloud Layers

The main gases in Jupiter's atmosphere are hydrogen and helium. There is also a little methane. The pale cloud bands seem to be made up of icy crystals of ammonia. The dark cloud bands seem to be made up of compounds of ammonia and sulphur. They are lower than the ammonia clouds.

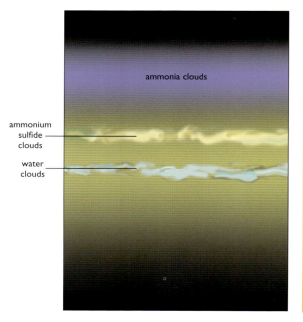

Above: The colourful face of Jupiter is crossed by dark belts and light zones. Dark and light spots show where there are stormy regions.

Left: Cloud layers in Jupiter's atmosphere.

Between the the main cloud layers there seems to be a haze of water-ice crystals. They are similar to the icy cirrus clouds on Earth, which are sometimes called mares' tails.

Furious Winds

The winds within the belts and zones travel at different speeds. They blow fastest around the equator and slow down to the north and south. Around the equator they can reach speeds of up to about 350 miles (560 km) an hour, more than the wind speed in a tornado on Earth.

The winds in the belts and zones also travel in different directions. Some blow toward the east, others toward the west.

JUPITER DATA

Diameter: 88,850 miles (142,980 km)

Average distance from Sun: 483,000,000 miles (778,000,000 km)

Mass (Earth=1): 318

Density (water=1): 1.3

Spins on axis in: 9.93 days

Circles around Sun in: 11.9 years

Number of moons: 16

in Jupiter's southern hemisphere. It is three times as big across as Earth and is known as the Great Red Spot. Astronomers have observed it for more than 300 years. It seems to be a region of high pressure in which spiraling winds carry gases high above the usual cloud layers. Its colour seems to come from compounds of phosphorus, a chemical element found on Earth that is often used in manufacturing matches.

Stormy Weather

Where the winds moving west meet the winds moving east, the result is an even greater disturbance, or turbulence, in the atmosphere. As a result of this turbulence, great waves form in Jupiter's atmosphere. Furious hurricane like storms break out. They appear to observers on Earth as pale and dark ovals.

The biggest one is a huge red oval region found in one of the cloud bands

Above: Furious winds swirl around in Jupiter's Great Red Spot, which measures more than 25,000 miles (40,000 km) across.

Below: Makeup of Jupiter's cloudy atmosphere. It is mainly hydrogen and helium.

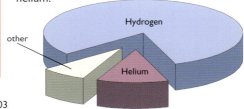

203

Magnetic Jupiter

As we saw earlier, Jupiter is made up mainly of hydrogen in three different forms. The atmosphere contains hydrogen gas. Beneath this there are deep layers of hydrogen in the form of a liquid and then in the form of a liquid metal.

All of Jupiter spins around very rapidly — its atmosphere, its liquid hydrogen ocean, and its liquid metal layer. But when metal moves, electric currents are created. These electric currents generate the force we call magnetism.

So, as the liquid metal in Jupiter moves with the planet's rotation, electric currents are created inside it, and giving the planet its magnetic field. Earth gets its magnetism in a similar way, because of electric currents set up in its metal core.

Jupiter's magnetism is much more powerful than Earth's and reaches out

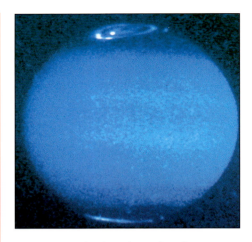

Above: Particles from Jupiter's radiation belts spill into its atmosphere and create auroras—glowing light displays around the north and south poles.

Below: Jupiter's magnetism shields the planet from the solar wind.

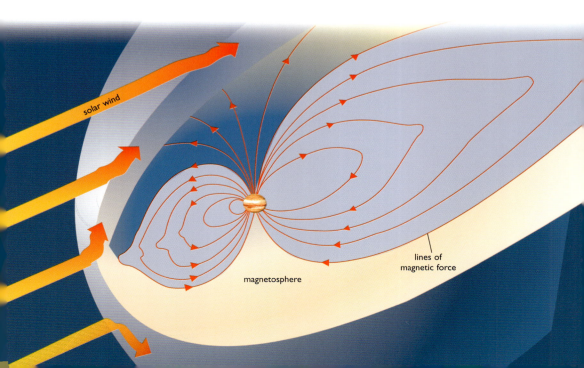

solar wind

lines of magnetic force

magnetosphere

millions of miles into space. It traps tiny electrical particles that pour into space from the Sun to form regions, or belts, of radiation. Earth has similar regions, called the Van Allen belts. The radiation from Jupiter's belts pose a great danger to spacecraft and could be deadly to any human crew.

Rings Around Jupiter

The Voyager space probes discovered many new things about Jupiter when they flew past the planet in 1979. One of their most unexpected discoveries was that Jupiter has a set of rings circling its equator it. They are much fainter than Saturn's famous rings and cannot be seen from Earth.

Jupiter's main ring is located about 40,000 miles (64,000 km) from the planet's cloud tops. This ring circles Jupiter around the orbits of the two nearest of the planet's moons — tiny Adrastea and Metis.

The main ring measures about 4,300 miles (7,000 km) across, but it is less than 20 miles (32 km) thick. It is made up of very tiny grains of dust, something like the smoke from bonfires on Earth.

Other rings are found both inside and outside the main ring. The inner ring is much fainter, but thicker, and is known as the halo. The outer ring is even fainter and is called the gossamer ring. Two more moons — Amalthea and Thebe — orbit inside the ring. Astronomers believe that the ring is made up of particles chipped off the two moons.

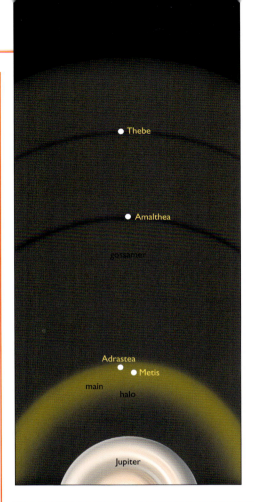

Jupiter's rings (right) are too faint to be seen from Earth. The very faint gossamer ring extends far beyond the main one, even beyond the orbits of the moons Amalthea and Thebe.

Many Moons

In January 1610, the Italian astronomer Galileo trained his newly built telescope on Jupiter. He saw little points of light like stars lined up on either side of the planet. These "stars" changed their position from night to night. He realized that they were tiny satellites, or moons, of Jupiter, circling in orbit around it. There were four moons in all.

These four moons can be seen through binoculars or a small telescope. We call them the Galilean moons in honour of their discoverer. In order of distance from Jupiter, they are Io, Europa, Ganymede, and Callisto.

Over the centuries, using telescopes astronomers discovered another nine smaller moons farther away from the planet. The Voyager space probes found three more, making a total of 16 moons in all.

Satellite Groups

The 16 moons of Jupiter divide neatly into three groups, widely separated from each other. The four Galilean moons form part of the inner group of eight moons. The other four in this group are much smaller than Io and lie closer to Jupiter. Callisto, the outermost of the inner group, orbits about 1,200,000 miles (1,900,000 km) from Jupiter.

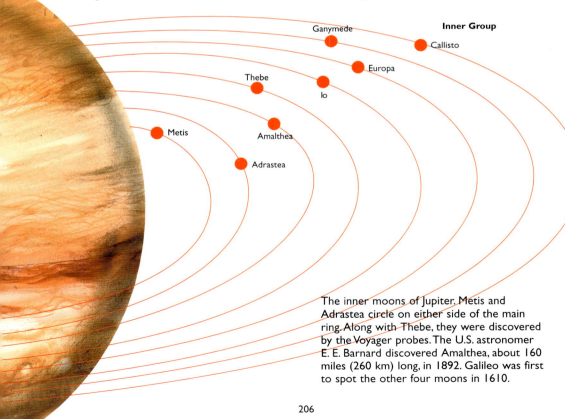

The inner moons of Jupiter. Metis and Adrastea circle on either side of the main ring. Along with Thebe, they were discovered by the Voyager probes. The U.S. astronomer E. E. Barnard discovered Amalthea, about 160 miles (260 km) long, in 1892. Galileo was first to spot the other four moons in 1610.

Earth

Moon

Europa

Io

Callisto

Ganymede

Above: Jupiter's four Galilean moons, compared in size with the Earth and our own Moon.

Leda

Himalia

Middle Group

Lysithea

Elara

Four moons make up a middle group, which is about seven times farther away from the planet. The innermost of these moons, called Leda, measures only about 6 miles (10 km) across. It is the smallest of all Jupiter's moons.

The outer group of four moons is twice as far away as the middle group. They orbit around Jupiter in the opposite direction of the others. Scientists say they have retrograde motion. The outermost of these moons, called Sinope, wanders as far as 15,000,000 miles (24,000,000 km) from Jupiter. It takes over two years to travel around the planet.

The Trojans

Jupiter's powerful gravity holds its family of moons. The planet's gravitational field also captured two groups of asteroids from the nearby asteroid belt. Both groups circle the Sun in the same orbit as Jupiter, one in front of the planet and one behind.

The middle group of Jupiter's moons, which all orbit more than 7 million miles (11 million km) from the planet. Tiny Leda was not discovered by telescope until 1974. Of this group, Himalia (105 miles, 170 km across) is the biggest.

These asteroid groups are called the Trojans. Each has been named after figures from the Trojan War, which the ancient Greeks fought against Troy, a city-state in what is now Turkey. The two largest are Hector, which is about 120 miles (190 km) in width, and Agamemnon, which is slightly smaller.

207

The Galilean Moons

The four Galilean moons dwarf Jupiter's other 12 moons and most other moons in the solar system. With a diameter of 3,273 miles (5,268 km), Ganymede is bigger even than the planet Mercury. Callisto is about the same size as Mercury, while Io is a little larger and Europa a little smaller than our own Moon.

Although they are relatively close together in space and are similar in size, the Galilean moons differ in composition and appearance.

Makeup of the Moons

Scientists think that Io has a solid core that is rich in iron. Io's core is surrounded by a thick layer of molten rock. On top is a thin, hard, rocky outer layer, or crust.

Europa also has an iron-rich core, surrounded by a thick layer of solid rock. Above this there is probably a deep ocean of water, with a layer of ice on top.

Ganymede also has an icy crust, with thick layers of ice and rock underneath. In the centre, there is an iron core, which may be partly molten.

Most of Callisto consists of a solid mixture of rock and ice, with no distinct layers. Its surface is icy, too.

The Ice Moons

Europa, Ganymede, and Callisto all have an icy surface but look quite different. Europa is amazingly smooth and bright — indeed, it is one of the

Below: The probable makeup of the four Galilean moons. They are all different. Much of Io is made up of molten rock. Europa may have an ocean of water under its surface.

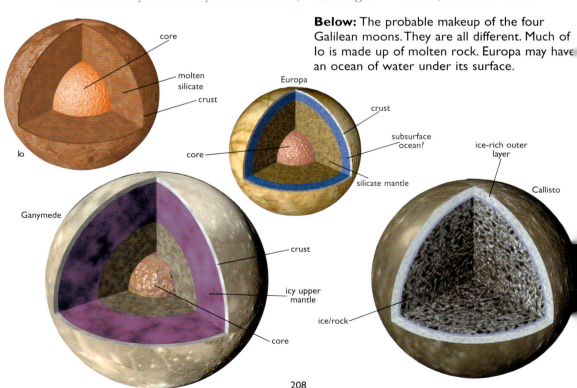

core

molten silicate

crust

Io

Europa

crust

subsurface ocean?

core

silicate mantle

ice-rich outer layer

Callisto

Ganymede

crust

icy upper mantle

ice/rock

core

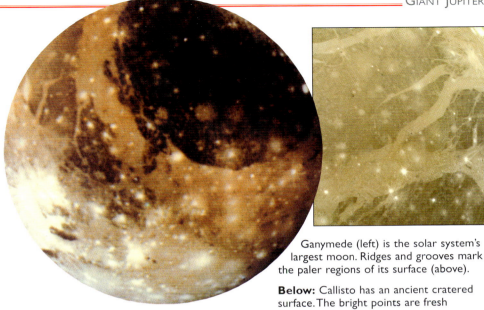

Ganymede (left) is the solar system's largest moon. Ridges and grooves mark the paler regions of its surface (above).

Below: Callisto has an ancient cratered surface. The bright points are fresh craters.

brightest bodies in the solar system. Ganymede is covered in dark and pale regions, while Callisto is dark all over.

Callisto's somber surface is pitted with craters, dug when meteorites or asteroids bombarded it long ago. The presence of so many craters on the surface indicates that Io must be very old. Here and there, white patches show where recent meteorite hits have thrown out fresh ice from craters.

Ganymede's surface is also marked with craters. There are more craters in the dark regions, which indicates that they are older than the brighter regions. The brighter regions are lighter in colour because they contain fresher ice than the darker regions. They are crisscrossed by sets of parallel ridges and valleys that are up to tens of miles wide and hundreds of miles long. This kind of feature is known as grooved terrain. It probably formed when the surface stretched and cracked.

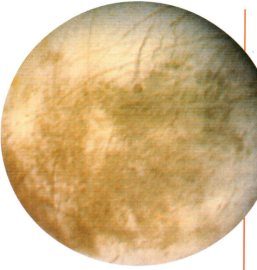

Europa has a smoother surface than any other moon.

Fascinating Europa

The Voyager space probes gave us our first views of the smallest Galilean moon, Europa. Images from the Voyager probe indicated that Europa has a very smooth icy surface, crisscrossed by a network of darker lines. The superior images sent back by a more recent probe, Galileo, have revealed a much more complex surface.

Galileo has shown that Europa has bright plains and darker mottled terrain, cut by networks of narrow ridges and grooves. These seem to be fractures, or breaks, in the icy surface. Scattered here and there are blocks of ice that appear to have broken free and drifted to new positions, like ice floes do in the Arctic Ocean on Earth.

Few craters are visible on Europa, suggesting that it has a very young surface. The craters that are present are not as prominent as those on the other moons. As meteorites hit the surface, the craters they make quickly fill with slushy material. They often appear darker than their surroundings because of material that wells up from under the icy surface.

In other parts of Europa domes and smooth areas known as puddles are found. These features could be caused by hot spots beneath the surface. Ultimately such sources could melt the ice beneath the surface and create an ocean of liquid water. Scientists have suggested that if there is an ocean of water, then Io's surface might contain some primitive forms of life.

Below: Fascinating patterns show up in the ice that forms the surface of Europa.

More than 80 active volcanoes have been spotted on Io, and there are many craters where volcanoes have erupted in the past. All Io's volcanoes are named after mythological fire gods and goddesses or other fiery subjects. The first volcano spotted on Io, Pele, was named for the legendary Hawaiian volcano goddess.

Left: In this Voyager picture, Io (left) and Europa are seen with Jupiter in the background.

Amazing Io

Io has the most unusual surface of all the Galilean moons. Io's surface is not dull and drab or ice-covered but vividly coloured in shades of yellow, orange, black, and white. We see little of the hard outer rocky crust of Io because it is covered almost entirely in the chemical element sulphur. Sulphur can be found in a variety of yellow and orange forms.

Where does this sulphur come from? On Io, it comes from erupting volcanoes. These volcanoes pour out molten sulphur and some molten rock as well. On Earth, volcanoes pour out mainly molten rock.

Earth volcanoes also give off gases, including sulphur dioxide. On Io, the volcanoes give off huge volumes of sulphur dioxide. In Io's low gravity, the gas shoots high above the surface, creating beautiful fountain-like streams, or plumes. The plumes we see in photographs are not of the gas itself but of sulphur dioxide snow — sulphur dioxide that turns into icy crystals in the cold vacuum of space.

Volcanoes are always erupting on Io, one of the most geologically active bodies in the solar system.

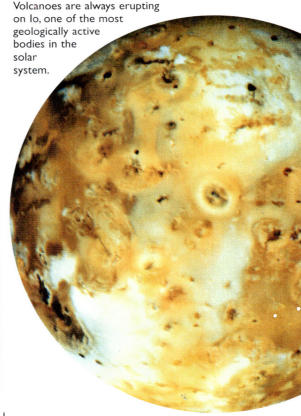

STUNNING SATURN

Through a telescope, Saturn appears perhaps the most beautiful planet of all because of the brilliant rings that circle it. It is the second-largest planet in the solar system, about four-fifths the size of Jupiter and nearly 10 times greater in diameter than Earth.

Saturn circles the Sun at an average distance of about 890 million miles (1,430 million km). It is the most distant planet that can be easily seen in the night sky with the naked eye.

From Earth, Saturn appears as a yellowish star. It never shines as bright in the night sky as Jupiter because it is much farther away — it never gets closer to Earth than about 800 million miles (1,300 million km). Its brightness in the sky varies greatly. Sometimes it shines more brilliantly than all the stars in the sky except Sirius and Canopus. But at other times it looks quite dim and is difficult to find among the true stars.

In make-up, Saturn is much like Jupiter, with an atmosphere of hydrogen and helium. Underneath, there are layers of hydrogen in the form of a liquid and in the form of a liquid metal. However, Saturn is much lighter for its size than Jupiter — it has a much lower density. Indeed, its density is lower than that of water. So if you could drop Saturn into a large enough bowl of water, it would float. No other planet would do this.

The face, or disk, of Saturn we see in a telescope is not as colourful as Jupiter's. We see similar bands of clouds travelling parallel to the planet's equator, but they are much fainter. The most noticeable feature of the disk is the dark shadow cast by the rings, which circle Saturn's equator.

All four giant outer planets have rings, but only Saturn's can easily be seen from Earth. Saturn's rings appear brighter than other planets' rings because they are broader and denser and because they are made up of icy particles not dark dusty matter.

Left: The surface of Enceladus, one of Saturn's many moons. Like Jupiter's large moons, it seems to have an icy surface, pitted with craters.

Saturn with two of its moons visible. The
face of the planet shows bands of clouds,
but it is much hazier than Jupiter's face.
The three classic rings show up clearly
here — C, B, and A moving outward
from the planet.

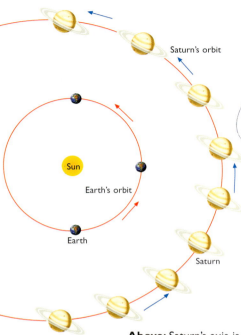

Saturn's orbit

Sun

Earth's orbit

Earth

Saturn

SATURN DATA

Diameter: 74,900 miles (120,540 km)
Average distance from Sun:
888,000,000 miles (1,429,000,000 km)
Mass (Earth=1): 95
Density (water=1): 0.7
Spins on axis in: 10.66 days
Circles around Sun in: 29.4 years
Number of moons: 18

Above: Saturn's axis is tilted in space. It always stays tilted in the same direction throughout its orbit around the Sun, which takes nearly 30 years to complete.

Spinning Around

Like all the planets, Saturn moves in two ways. It both travels in space in an elliptical orbit around the Sun and it spins around, or rotates, on its axis. The length of time that it takes to complete a single rotation is the length of Saturn's day.

Saturn spins around in space very rapidly, rotating once in about 10½ hours. Among the planets, only Jupiter spins around faster. This rapid spinning causes Saturn to bulge out noticeably at the equator and flatten out at the poles. The planet becomes misshapen in this way because it is made up of fluid (gas and liquid), and fluids can change their shape easily.

Magnetic Saturn

Like Jupiter, Saturn contains a thick layer of liquid hydrogen in the form of a

Because Saturn's axis is tilted, we get different views of the rings year by year as we look at it from Earth. Sometimes we see the rings straight on (1). Over time, we see the rings open up. Gradually they close up until after nearly 15 years, we see them straight on again (7).

metal. As this metal layer spins around with the planet, it creates electrical currents and magnetism. Saturn does not have as strong a metallic pull as Jupiter.

Saturn's Year

Saturn takes nearly 30 Earth-years to circle once around the Sun. During this time we get different views of the planet because of the way it spins around in space. It does not spin in an upright position as it circles the Sun. It spins around at an angle — its axis in space is tilted. (Earth's axis is tilted in space in a similar way.)

Because of the tilt of its axis, we view Saturn at a slightly different angle every year. In particular, we see slightly different views of its rings. When the planet's axis is tilted most toward us, we get the clearest view of the rings' circular openings. Saturn's axis is tilted in a different direction, we see the rings from the side, which makes them almost invisible.

Atmosphere and Weather

Saturn has a hazy, cloudy atmosphere. The clouds speed around the planet in parallel bands. The darker bands are called "belts," and the lighter ones, "zones." Within the bands, the winds blow mainly toward the east, following the planet's direction of rotation. In some bands the winds blow in the opposite direction, toward the west.

The strongest winds blow around the equator, where they can reach speeds of more than 1,100 mph (1,800 km/h). This is about five times the speed of the strongest hurricane winds on Earth.

Where winds blowing toward the east meet winds blowing toward the west, the atmosphere experiences turbulence. Great hurricane-like storms are formed.

From Earth they appear as pale, dark patches or spots. Some of these features last for years. One discovered by the Voyager probes, named Anne's Spot, was deep red, like Jupiter's Great Red Spot but much smaller.

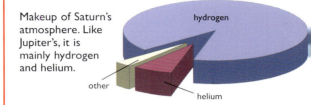

Makeup of Saturn's atmosphere. Like Jupiter's, it is mainly hydrogen and helium.

Below: Cloud layers in Saturn's atmosphere.

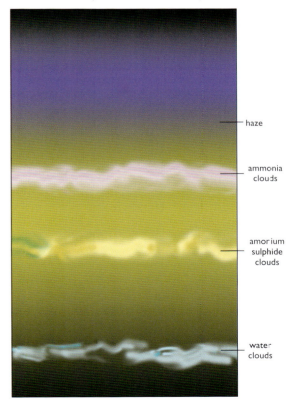

haze

ammonia clouds

amorium sulphide clouds

water clouds

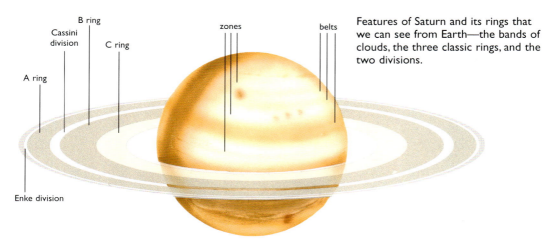

B ring

Cassini division

C ring

A ring

zones

belts

Enke division

Features of Saturn and its rings that we can see from Earth—the bands of clouds, the three classic rings, and the two divisions.

The Glorious Rings

When the Italian astronomer Galileo first observed Saturn through a telescope in 1610, he noticed that there was something strange about the planet. The main body seemed to have "companions" on either side. What Galileo was seeing was the rings extending on each side of the planet, but his telescope was not powerful enough to show them as rings.

Later astronomers saw the rings more clearly but thought that they were solid. In 1875 the English scientist James Clerk Maxwell proved that solid rings could not exist because they would be torn apart by the forces set up by Saturn's enormous gravity. The rings had to be made up of separate lumps of material.

Simple as ABC

Through telescopes, Saturn appears to have three rings, named A (the outer), B, and C (the inner). The B ring is the

Right: Closer-up, several more rings become visible. The very faint outer E ring surrounds the orbits of several of Saturn's large moons.

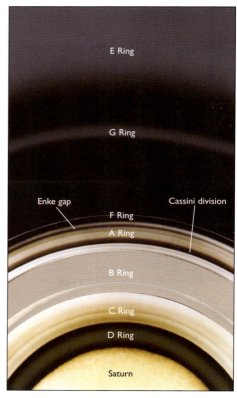

E Ring

G Ring

Enke gap

Cassini division

F Ring

A Ring

B Ring

C Ring

D Ring

Saturn

brightest of the rings. It contains more matter than the other rings, which makes it reflect sunlight better. After the B ring, the A ring is brightest. The C ring is much fainter and contains so little matter that it is transparent.

From one side to the other, the visible ring system measures about 170,000 miles (274,000 km) across. The B ring is the widest of the rings, measuring some 15,000 miles (25,000 km) across. It is separated from the A ring by a dark gap, which is called the Cassini Division. There is also a narrower gap near the outer edge of the A ring, called the Encke Division.

New Rings and Ringlets

When the Pioneer 11 and the two Voyager probes visited Saturn, they discovered several more rings. They found an even fainter ring inside the C ring. This D ring probably extends right down to Saturn's cloud tops.

The other rings lie outside the visible ring system. Close to the A ring lies a very narrow F ring. Farther out lie a broader but fainter G ring and beyond that a very broad and very faint E ring. The probes also showed that the three rings we see from Earth are made up of thousands upon thousands of separate ringlets.

One of the most beautiful images of Saturn and its rings sent back by the Voyager probes. The B ring shines brightest here. Like the other rings, it is made up of thousands of narrow ringlets.

217

Racing Around

The thousands of ringlets that together form Saturn's rings are made up of bits of matter that travel around the planet at high speed. These bits seem to be made up mainly of water ice, which is why the rings reflect light so well. The bits vary widely in size — some are smaller than pebbles, while others are as big as boulders, up to 33 feet (10 metres) across.

Below: By computer processing Voyager images into false-colour pictures, the differences in particle sizes in the rings can be seen.

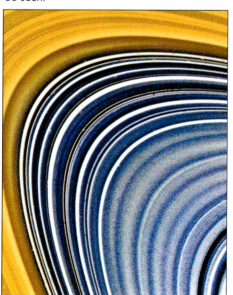

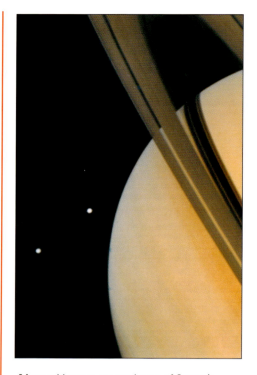

Above: Voyager spotted two of Saturn's moons close to the planet's limb. They are Tethys (top) and Dione, which are roughly the same size — about 700 miles (1,100 km) across.

The Shepherds

Several of Saturn's many moons are found within the system of rings. The relatively large satellites Mimas and Enceladus, for example, circle the planet within the outer E ring. Of even greater interest are three tiny moons found on either side of the narrow F ring.

The smallest one circles close to the outer edge of the A ring. Named Atlas, it is no more than about 11 miles (18 km) in diameter. Astronomers believe that this tiny moon somehow helps keep the particles in the A ring in place. They call it a shepherd moon because it seems to "herd" the ring particles just as a shepherd herds a flock of sheep.

In a similar way, two small moons seem to keep the particles in the narrow F ring in place. These shepherd moons, named Pandora and Prometheus, orbit on either side of the ring.

Many More Moons

The three "shepherds" mentioned were among several new moons discovered by the Voyager probes. The probe found other small moons circling in the same orbits as some of the large moons that we can see from Earth, such as Tethys and Dione. Altogether Saturn has at least 18 known moons, and others have been glimpsed from time to time but not confirmed by repeated sightings.

Tethys and Dione form part of a group of inner moons around Saturn, which also includes Mimas and Enceladus, closer in, and Rhea, farther out. More than twice as far away from Saturn as Rhea are a pair of moons, vastly different in size. First comes Titan, more than 3,000 miles (5,000 km) in diameter, then Hyperion, which is only one-tenth Titan's size.

Far Apart

Saturn's two outer satellites are even more widely separated. Iapetus lies twice as far away from Saturn as Hyperion, while Phoebe lies nearly four times farther away.

Phoebe does not circle around Saturn in the same direction as the other moons. It circles in a clockwise direction. Astronomers say it has a retrograde orbit.

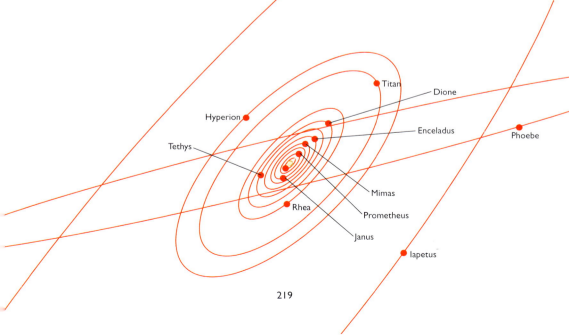

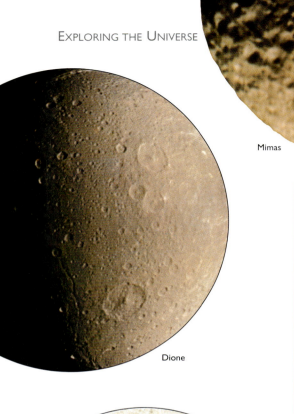

Mimas

Dione

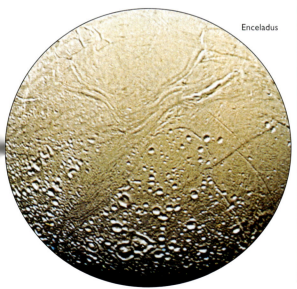

Enceladus

Icy Moons

The tiny moons of Saturn that space probes discovered are irregularly shaped lumps no bigger than about 130 miles (220 km) in diameter. They are probably made up of rock.

Most of Saturn's moons that can be seen from Earth are spherical, or ball-shaped. Most seem to be made up of a mixture of ice and rock, rather like the large moons of Jupiter. Most have an icy surface. The exception is Titan, which is very different from the others.

Saturn's icy moons might all be similar in make-up, but they all look different from one another. Mimas is the nearest large moon to Saturn. It is heavily cratered. Its largest crater, 80 miles (130 km) across, is named Herschel in honour of the English astronomer William Herschel, who discovered the moon in 1789.

Farther out and a little larger than Mimas is Enceladus. Its surface contains many large smooth regions. In places they are crossed by what look like ridges and valleys. These features were probably formed when the icy crust moved and cracked. It has fewer craters than Mimas, and they appear as though they formed quite recently. This suggests than Enceladus has quite a young surface that may still be changing.

The next two icy moons beyond Enceladus are Tethys and Dione, which are roughly the same size. Next comes

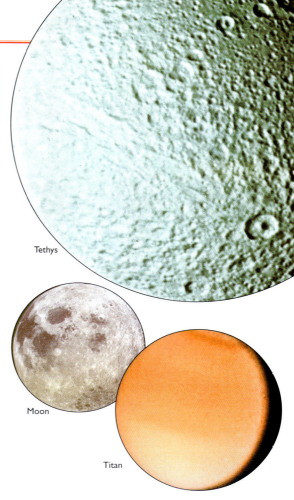

Left: Mimas was the first of Saturn's moons to be found, by William Herschel in 1789, eight years after he had discovered the planet Uranus. It measures about 250 miles (400 km) across.

Right: Tethys is heavily cratered and also has a long, deep valley cutting across it. You can see it in the picture, ending just before the prominent crater (Telemachus) near the bottom. The valley, named Ithaca Chasma, is about 1,200 miles (2,000 km) long.

Tethys

Moon

Titan

Rhea, which is 50 percent larger. All three moons are peppered with craters, although they also have smoother plains regions as well.

The most notable feature on Tethys is a huge valley system, known as Ithaca Chasma. Some 1,200 miles (2,000 km) long, it stretches three-quarters of the way around the moon.

Titan

Titan is by far Saturn's largest moon, with a diameter of 3,200 miles (5,150 km). This makes it larger than any other moon in the solar system except for Jupiter's Ganymede. It is bigger than the planet Mercury.

Astronomers think that Titan is made up of a mixture of rock and ice, like Saturn's other large moons. But it is much denser than the other moons, which means it contains more rocky matter.

We don't know what the surface of Titan looks like because the moon has a thick atmosphere. It is the only moon in the solar system to have one. The main gas in the atmosphere is nitrogen, which is also the main gas in Earth's atmosphere. Other gases in the atmosphere include methane (the gas we use on Earth for cooking and heating).

Above: Titan is much bigger than our Moon. Unlike our Moon, it has a thick atmosphere.

Methane is just one of several carbon compounds found in the atmosphere. When these carbon compounds break down in the atmosphere, they create an orange smog.

The temperature on Titan is about −180°C. At such a temperature, methane and other carbon gases may turn into liquid and form pools or even oceans on the surface. They may also freeze solid, to form snow and ice.

URANUS AND NEPTUNE

Far beyond Saturn in the depths of the solar system lie the planets Uranus and Neptune, which were unknown to ancient astronomers. Like Saturn, they are made up mainly of gas and liquid. They are much smaller than Saturn but are still about four times as great in diameter as Earth.

Ancient astronomers knew of only six planets, including Earth. They could see the other five planets moving around the heavens like wandering stars. The most distant and faintest of them was Saturn.

In 1781 an English astronomer named William Herschel found a seventh planet. In March of that year he was looking at the stars in the constellation Gemini and spied an object he had not noticed before. He knew it wasn't an ordinary star because it appeared as a disk — stars always appear as a point.

Herschel wrote in his notebook that the object was "a curious either nebulous star or a comet" and came to the conclusion that it must be a comet. But he was wrong. Later, astronomers calculated its orbit and proved that it must be a new planet that circled the Sun twice as far away as Saturn. At a stroke, the size of the solar system had been doubled. The new planet was eventually named Uranus, after one of the oldest Greek gods.

As astronomers began plotting the path of Uranus through the heavens, they found that it did not keep to the orbit they calculated it should follow. Something appeared to be throwing it off course. Astronomers believed that yet another undiscovered planet was to blame, pulling Uranus off course with its gravity.

In 1845, an English mathematician named John Couch Adams calculated where this new planet should be. Even so, he could not persuade any astronomers to look for it. Meanwhile, a French mathematician, Urbain Leverrier, had also worked out the position of the new planet. In September 1846, a German astronomer, Johann Galle, at Berlin Observatory found the planet almost exactly where Leverrier (and Adams) had predicted. The new planet was named Neptune, after the Roman god of the sea.

Opposite: A large stormy region on Neptune spied by Voyager 2. Called the Great Dark Spot, it is about the size of Earth. It is edged by wispy clouds of frozen methane.

Looking at Uranus

Uranus is the seventh planet in distance from the Sun, and the third largest. With a diameter of 31,765 miles (51,120 km), it is slightly larger than Neptune. At its closest distance to Earth (about 1.7 billion miles, 2.8 billion km), it can be glimpsed with the naked eye as a very faint star, if you know where in the sky to look for it.

Telescopes show Uranus as a pale greenish-blue disk, with no definite markings. Even close-up pictures of the planet taken by the Voyager 2 probe in 1986 reveal few features. There are no signs of the bands of clouds seen on Jupiter and Saturn.

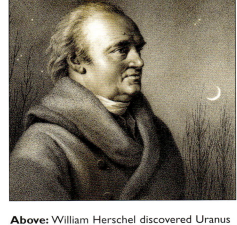

Above: William Herschel discovered Uranus in 1781. He also discovered its two largest moons, Titania and Oberon, six years later.

Above: No features can be seen on the face of Uranus. The methane clouds in the atmosphere are hidden by a thick haze.

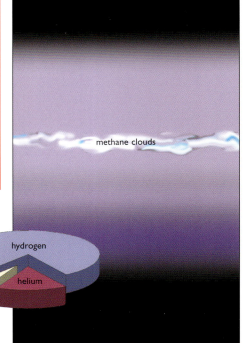

methane clouds

hydrogen

methane

helium

Right: Makeup of Uranus's atmosphere. Hydrogen, helium, and methane are the main gases.

Like most planets, Uranus spins around an axis that is tilted in space as it orbits the Sun. Neptune has a tilted axis, too. Neptune's axis is only slightly tilted, but Uranus's axis tilts so far that the planet travels on its side.

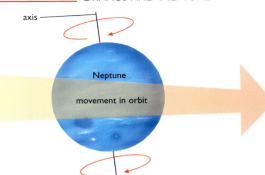

axis

Neptune

movement in orbit

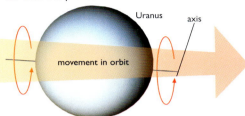

Uranus

axis

movement in orbit

Uranus is a fluid (gas and liquid) planet like Jupiter and Saturn but is different in makeup. It has a very deep atmosphere of hydrogen and helium, with a certain amount of methane. It is the methane that gives the atmosphere its blue-green colour.

Beneath the atmosphere, Uranus is covered by a deep, warm ocean that contains water, ammonia, and methane. At its centre is a rocky core. Currents swirling in the ocean as the planet spins around create electric currents that make the planet magnetic. Uranus's magnetic force is about as strong as Earth's.

Uranus in Motion

Uranus is so far away that it takes nearly 84 Earth-years to travel once around the Sun. Like the other planets, it spins around in space on its axis, turning around once in about 17 hours.

Most planets rotate in a more or less upright position as they orbit the Sun. But Uranus is different. It spins around on its side. This means that its north and south poles take turns facing the Sun. We do not know why Uranus spins on its side. Maybe the planet was knocked into this position in a collision with another body long ago.

New Moons

From Earth, we can see five large moons circling around Uranus. William Herschel discovered the first two — Oberon and Titania — in 1797. The others are Miranda, Ariel, and Umbriel.

Voyager 2 took close-up pictures of these moons when it flew past Uranus in 1986. During this mission it also discovered 12 more tiny moons orbiting closer to the planet. The inner two, Cordelia and Ophelia, are the tiniest, measuring only about 15–20 miles (25–30 km) in diameter.

URANUS DATA

Diameter: 31,765 miles (51,120 km)
Average distance from Sun:
1,787,000,000 miles (2,875,000,000 km)
Mass (Earth=1): 15
Density (water=1): 1.3
Spins on axis in: 17.24 days
Circles around Sun in: 83.7 years
Number of moons: 18

225

Miranda

Umbriel

The five large moons of Uranus, in order of distance from the planet. Their sizes are shown to scale. Smallest is Miranda, which is only about 300 miles (485 km) in diameter.

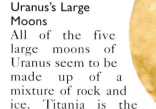

Titania

Uranus's Large Moons

All of the five large moons of Uranus seem to be made up of a mixture of rock and ice. Titania is the largest, with a diameter of about 980 miles (1,580 km). Its icy surface is pockmarked with craters that measure up to 125 miles (200 km) across. It is crossed by valleys, which are faults, or cracks, in the surface. One is more than

900 miles (1,500 km) long — more than three times as long as Arizona's famous Grand Canyon.

The slightly smaller Oberon is more heavily cratered, and many of the craters contain dark material. Umbriel is also well cratered and is darker than the other large moons. With a diameter of about 730 miles (1,170 km), it is about the same size as Ariel. Ariel has many craters and networks of deep valleys. It also has large regions of fresh ice, which make it the brightest of the moons.

Miranda, which is about half the size of Ariel, has the most unusual surface of all. It has very different kinds of landscape mixed together, like the different patches in a patchwork quilt. Some regions seem ancient and heavily cratered. Some have strange curving grooves. In other regions ice cliffs soar more than 12 miles (20 km) high.

Rings Around Uranus

Until 1977, Saturn was the only planet known to have rings. In March of that year astronomers accidentally discovered that Uranus also has rings. They were trying to measure the size of the planet

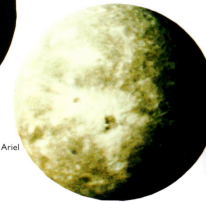

Oberon

Ariel

accurately by timing how long it took to pass across a distant star.

But just before the star passed behind Uranus, it "winked" several times, showing that something had passed in front of it. The star "winked" again just after it came from behind the planet. Other astronomers carried out similar experiments with the same results — "winks" before and after stars passed behind Uranus. They realized that the "winks" were being caused by sets of rings around the planet.

Voyager 2 pictured the rings clearly in 1986. There are 11 rings in all — 10 narrow and relatively bright ones and, inside them, 1 that is very broad but very faint. The rings seem mostly to be made up of lumps of dark material as much as 33 feet (10 metres) in width.

The brightest ring is the outermost one, called the Epsilon. It circles about 15,000 miles (25,000 km) above Uranus's atmosphere. It varies in width between about 12 and 60 miles (20 and 96 km). The other main rings are much narrower, some less than a mile across. The tiny moons Cordelia and Ophelia orbit on either side of the Epsilon ring and seem to act as "shepherds" to keep the ring particles in place.

Above: An artist's impression of Uranus and its rings. In reality, they are dark and difficult to make out.

Below: Voyager 2 took this close-up picture of Uranus's rings, showing fine dust particles in between the ringlets.

NEPTUNE DATA

Diameter: 30,779 miles (49,530 km)

Average distance from Sun:
2,799,000,000 miles (4,504,000,000 km)

Mass (Earth=1): 17

Density (water=1): 1.6

Spins on axis in: 16.11 days

Circles around Sun in: 163.7 years

Number of moons: 8

Makeup of Neptune's atmosphere. The main gases are hydrogen and helium, with traces of methane.

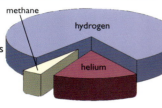

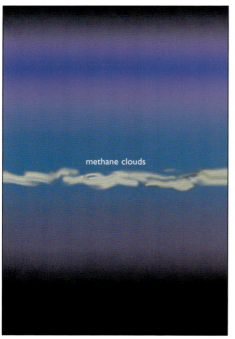

methane clouds

Looking at Neptune

Neptune is a little smaller than Uranus but has much the same makeup, with a thick atmosphere and a warm, deep liquid ocean underneath. It lies about a billion miles (1.6 billion km) farther away from the Sun than Uranus and takes nearly 164 Earth-years to circle the Sun once.

Strangely, even though it is much farther away from the Sun than Uranus, Neptune has more or less the same temperature, about –210°C. This means that Neptune is somehow being heated from within. This internal heat produces much more weather in the atmosphere than we would expect for a planet so far away from the Sun.

Stormy Weather

The main features of Neptune's weather are clouds, winds, and storms. The clouds are white and wispy, like the high cirrus clouds we find on Earth. They are probably made up of crystals of frozen methane.

Winds surge around the planet at high speed, in places reaching 1,500 mph (2,400 km/h) or more. This is five times the speed of winds in the most violent twisters, or tornadoes, on Earth. In places the winds swirl around to form hurricane-like storms. Voyager 2 spotted several dark oval regions where violent storms were raging. The largest was named the Great Dark Spot. These storm regions had disappeared by the 1990s when the Hubble Space Telescope began photographing the planet.

Methane clouds float in Neptune's atmosphere. The upper atmosphere is much less hazy than on Uranus, giving the planet a much deeper blue colour.

Like earth, Neptune is a blue planet. On Earth, the colour comes from the blue of the oceans. On Neptune, the colour comes from the methane in the atmosphere.

Neptune's Moons

The little shepherds are among the six new moons of Neptune found by Voyager 2. They seem to be dark, shapeless bodies. Voyager also flew close to Triton, one of the two moons that can be seen from Earth. Triton is a little smaller than our own Moon but is quite a different body. It is a deep-frozen body covered with frozen nitrogen and methane gas. In places icy geysers erupt, shooting fountains of gas and ice high above the surface.

Neptune's Rings

After finding rings around Uranus, astronomers began to suspect that Neptune might also have rings around it. This proved to be true. The rings were first pictured clearly by Voyager 2, when it visited the planet in 1989.

There are four rings in all — two narrow and bright ones and two that are much broader and fainter. The two bright ones are named Adams and Leverrier after the mathematicians who played key roles in the planet's discovery. The bright Adams is the outermost ring, circling the planet at a distance of about 24,000 miles (40,000 km).

Its ring particles seem to be kept in place by a tiny shepherd moon, named Larissa. Two of the other rings have tiny shepherds, too.

Neptune's rings are named after people who played a part in the planet's discovery. Johann Galle, for example, was the astronomer who first spotted Neptune on September 23, 1846.

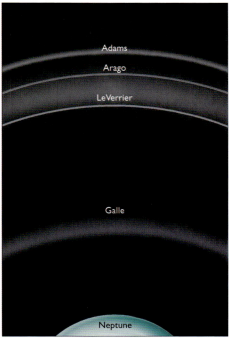

229

PLUTO AND CHARON

Pluto was the last planet to be discovered. It is the smallest planet by far. Pluto is only about two-thirds the size of our Moon. It is quite different in makeup from all the other planets, consisting mainly of water, ice, and rock. Its only moon, Charon, is half its size.

Neptune was discovered in 1846 after astronomers had found that Uranus was not following the orbit it should. Later, astronomers found that both these planets still orbited in a slightly strange way. So they began to think that there might be yet another planet affecting them.

The first attempts to find a ninth planet were made in the 1870s, but it was not until 1905 that a systematic search began. It was led by the U.S. astronomer Percival Lowell at Flagstaff Observatory in Arizona. Lowell had built the observatory originally to study Mars. The search for the unknown planet, which Lowell called Planet X, revealed nothing by the time of his death in 1916.

It was not until 1929 that the search resumed at Lowell Observatory. A young astronomer from Kansas named Clyde Tombaugh was asked to make and examine photographic plates for new objects in the region where Lowell had thought Planet X might be.

In February 1930, Tombaugh was examining plates he had taken a few weeks earlier and spotted what he was looking for. It was a body far beyond Neptune and much fainter than expected. It was a ninth planet, which came to be called Pluto, after the god of the dark underworld of the dead in Roman mythology.

Astronomers were not able to find out much about Pluto — not even its size — until the 1970s. In 1978, U.S. astronomer James Christy discovered that Pluto had a moon, which was called Charon. By observing the way the moon circled Pluto and using basic laws of motion, astronomers could at last work out Pluto's size and mass. It proved to be just 1,430 miles (2,300 km) in diameter, much smaller than our own Moon. Charon proved to be only half Pluto's size.

The strange thing about the story of Pluto's discovery is that the planet is much too small to have any gravitational effect on Neptune and Uranus.

In recent times Pluto has been downgraded from being called a planet as scientists now believe that it could actually be an asteroid.

Opposite: Pluto and its moon, Charon, pictured at the edge of the solar system. Here we see them as crescents, lit up faintly by the distant Sun, nearly 4 billion miles (6.4 billion km) away.

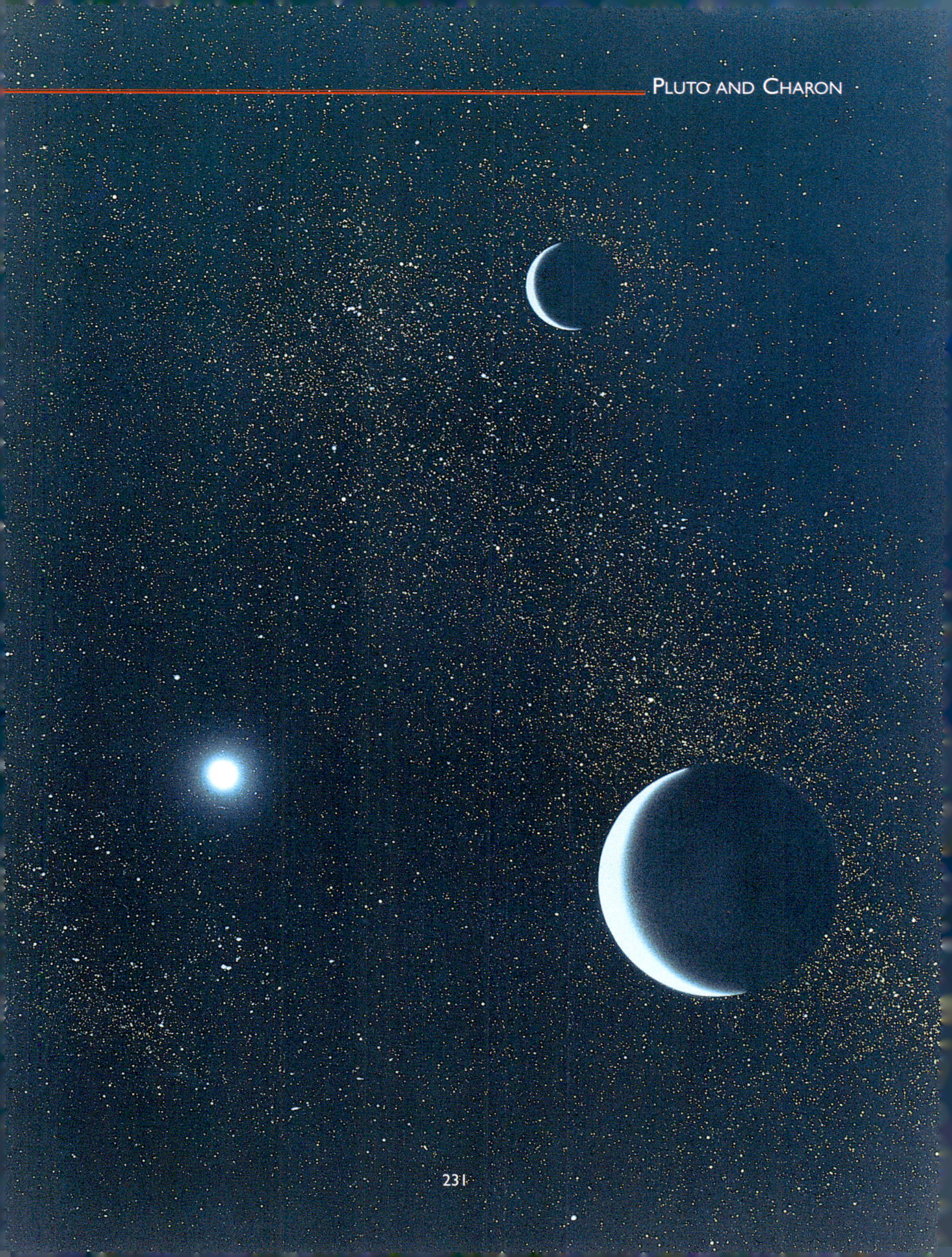

The Farthest Planet?

Tiny Pluto is the planet that wanders farthest from the Sun. We cannot say simply that Pluto is the farthest planet, because at times it is not. Between 1979 and 1999, Pluto was actually closer to the Sun than Neptune, and during that time Neptune was the farthest planet.

Pluto wanders inside Neptune's orbit for 20 years of the 248 years it takes to circle once around the Sun. Pluto's orbit is highly elliptical, or oval. As a result, its distance from the Sun varies widely, between about 2.7 and 4.6 billion miles (4.4 and 7.4 billion km).

Pluto's orbit is unusual in another way. Most planets have orbits that lie more or less in the same plane, or flat sheet, in space. But Pluto's orbit reaches high above and below this plane.

Icy World

Pluto is the only planet in the solar system that has not been explored at all by space probes. This means that we know less about it than we do about the other planets.

Pluto is so far away that in telescopes on Earth it looks just like a star, and we can see nothing of its surface. Recently the Hubble Space Telescope has sent back pictures showing some vague light and dark patterns on its surface.

The surface seems to be covered with icy, frozen gases such as nitrogen and methane and even water ice. The light markings we see in the Hubble pictures are probably icy regions. The darker areas might be patches of exposed rock or they might be regions covered by reddish-brown chemical compounds.

Below: Pluto and Charon, photographed from space by the Hubble Space Telescope. No telescopes on the ground are powerful enough to show the two as separate bodies.

Above: We do not know what Pluto really looks like because no space probes have visited it yet. But it could look similar to some of the icy moons of Uranus and Neptune, such as Triton.

Pluto has the most eccentric, or most oval, orbit in the solar system. This orbit takes it way above and below the orbits of the other planets.

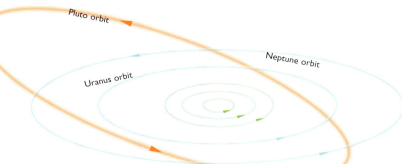

Pluto orbit

Neptune orbit

Uranus orbit

In the feeble heat from the Sun, the surface ice on Pluto slightly evaporates, or turns to gas. This gives the planet a very thin atmosphere. However, this gas turns back to ice as Pluto moves farther from the Sun.

Charon

Pluto's moon, Charon, is so big that Pluto-Charon is often called a double planet. It orbits close to Pluto, at an average distance of less than 12,000 miles (19.000 km). By comparison, our own Moon circles Earth 20 times farther away.

Pluto spins around on its axis once in a little over 6 days. Charon circles around Pluto in exactly the same amount of time. This means that the moon stays fixed in the same position in Pluto's sky.

Below: Both Pluto and Charon spin around axes that are tilted by the same amount. And both spin round these axes in the same direction, which is the opposite direction to most of the other planets. Pluto and Charon also spin round once in space in the same time — about 6.4 Earth-days.

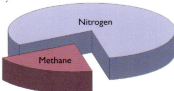

Nitrogen

Methane

Probable makeup of Pluto's very faint atmosphere.

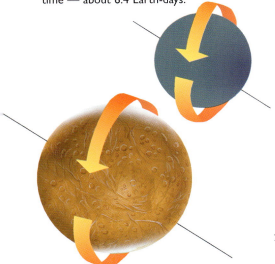

PLUTO DATA

Diameter: 1,429 miles (2,300 km)
Average distance from Sun: 3,676,000,000 miles (5,916,000,000 km)
Mass (Earth=1): 0.002
Density (water=1): 2.0
Spins on axis in: 6.4 days
Circles around Sun in: 248 years
Number of moons: 1

233

Timeline of Important Dates

3000 BC Astronomy well established in Middle East

585 BC Greek astronomer Miletus correctly forecasts an eclipse

150 BC About this time the Greek astronomer Hipparchus catalogs more than 1,000 stars and introduces the scale of magnitude for estimating star brightness.

AD 150 About this time Ptolemy writes an encyclopedia of the scientific and astronomical knowledge of the day, containing the idea of an Earth-centred universe.

800s Arab astronomy flourishes

1054 Chinese astronomers record a supernova in the constellations Taurus (Bull), which gave rise to the Crab Nebula we see today.

1543 Nicolaus Copernicus advances his idea of a solar system

1576 Tycho Brahe sets up advanced observatory on the isle of Hven

1609 Italian astronomer Galileo first observes the heavens in a telescope

1610 Galileo discovers sunspots

1668 Isaac Newton in Britain builds a reflecting telescope.

1675 Greenwich Observatory founded in England

1781 William Herschel discovers Uranus

1801 Giovanni Piazzi discovers the first asteroid, Ceres

1801 J. Ritter discovers that the Sun gives out ultraviolet radiation

1802 William Herschel discovers binary stars

1840 J. W. Draper takes the first photographs of the Moon

1908 George Hale first measures the Sun's magnetism

1912 US astronomer Henrietta Leavitt studies Cepheid variables and establishes a relationship between their true brightness and period (time in which they change brightness).

1917 100-inch Hooker reflector telescope completed at Mt. Wilson Observatory

1923 Edwin Hubble proves that galaxies are distant star systems

1931 US engineer Karl Jansky detects radio waves coming from the heavens and launches the science of radio astronomy.

1948 200-inch Hale reflector telescope completed at Mount Palomar Observatory

1957 Russia's *Sputnik 1* launches the Space Age

1959 The *Luna 2* probe crash-lands on the Moon

1963 Arecibo radio telescope completed

1966 *Surveyor 1* makes a soft landing on the Moon and transmits the first close-up photographs of its surface

1967 Radio astronomers at Cambridge, England, discover pulsars.

1977 *Voyager* probes launched to the outer planets

1981 Very Large Array radio telescope completed

1968 *Apollo 8* carries the first astronauts to the Moon and back, but doesn't land

1969 *Apollo 11* makes the first Moon landing

1972 The final Apollo mission (*Apollo 17*) takes place

1973 Astronauts on the U.S. space station Skylab make a detailed study of the Sun

1990 Hubble Space Telescope launched

1991 One of the longest solar eclipses of the century is observed from Hawaii

1995 The SOHO probe is launched to study the Sun

1997 NASA scientists discover fountains of antimatter near the centre of the Milky Way

1997 Launch of the *Cassini-Huygens* mission to

Jupiter and Saturn; it successfully arrived in July 2004

1997 *Asiasat 3* launched by the People's Republic of China

1998 John Glenn returns to space at age 77, 36 years after he orbited the Earth for the first time

1998 *Lunar Prospector* reached Moon in January

1998 The *Lunar Prospector* probe discovers ice in lunar craters

1998 United States launches *Deep Space 1*

1998 United States launches the *Mars Climate Orbiter* and *Mars Polar Lander*

1999 Building of a new international space station commences

1999 Eileen Collins aboard *Columbia* becomes the first woman to command a US space shuttle mission

1999 US launches Chandra X-ray Observatory, more powerful than the Hubble Space Telescope

1999 Sightings of planets around other stars

1999 United States launched *Deep Space 2*, the Martian Microprobe Project

1999 The Spacewatch team discover the 17th moon of Jupiter, Callirrhoe

1999 Stephano is the twentieth moon of Uranus to be found

2000 *NEAR-Shoemaker* goes into orbit around asteroid Eros

2000 A Russian and US crew resides on the International Space Station for four months

2000 Ten satellites of Jupiter are discovered

2000 Observers in La Silla and Mauna Kea discover numerous new moons of Saturn

2001 Scientists find evidence for a black hole at the centre of our galaxy

2001 Mir space station re-enters Earth atmosphere after more than 86,000 orbits

2001 Leonid meteor storm on November 18

2002 Observations from a spacecraft orbiting Mars suggest large deposits of ice may lie below the Martian surface.

2002 Observations suggest large deposits of ice may lie below the Martian surface

2003 Space shuttle Columbia breaks up during Earth re-entry, killing its 7 crew

2003 New studies in February indicate that the Universe is 13.7 billion years old

2003 On August 27, Mars made its closest approach to the Earth in more than 60,000 years

2003 The total of known Jovian moons reaches 60 in this year

2004 Evidence of icy water found on Mars by European Mars Orbiter craft

2004 Astronomers announce the discovery of a planetoid, Sedna, orbiting the Sun every 10,000 years, well beyond Pluto

2004 Cassini-Huygens probe arrives at Jupiter and Saturn, sending back a wealth of new information

2004 *Genesis* spacecraft brought pieces of the Sun from space

2004 The Cassini Imaging Science Team discover several new moons of Saturn

2005 Astronomers identified the three largest stars yet discovered: KW Sagitarii, V354 Cephei and KY Cygni each have radii about 1,500 times greater than that of the Sun

2005 Space Shuttle Discovery launches in July, the first new mission since the February 2003 *Challenger* explosion

2005 *Cassini* mission discovers seas of methane on Titan

2006 *Stardust* capsule successfully returns to Earth after 3 billion mile trip to gather cometary dust samples

Glossary

active galaxy A galaxy that gives out very much more energy than usual, often as radio waves or X-rays.

asteroids Small lumps of rock or metal that circle the Sun. Most circle in a broad band (the asteroid belt) between the orbits of Mars and Jupiter.

astrology A belief that people's characters and everyday lives are somehow affected by the stars and planets.

astronomy The scientific study of the heavens and the heavenly bodies.

atmosphere The layer of gases around a heavenly body.

atoms The smallest bits of a substance. Every atom has a centre, or nucleus, with electrons circling around it.

big bang A fantastic explosion that astronomers think created the Universe about 15 billion years ago.

binary A two-star system, in which two stars circle around each other, bound by gravity.

black hole A region of space with enormous gravity; not even light can escape from it.

blueshift A movement, or shift, in the lines in the spectrum of a star or galaxy toward the blue end. It indicates that the object is travelling toward us.

celestial sphere An imaginary dark globe that appears to surround the Earth. The stars seem to be fixed to the inside of the sphere.

cepheid A variable star that changes in brightness regularly over a few hours or a few days.

chasma A deep valley.

chromosphere The inner part of the Sun's atmosphere.

climate The average kind of weather a place experiences during the year.

cluster A group of stars or galaxies. See open cluster, globular cluster, supercluster.

comet A small icy lump that gives off clouds of gas and dust and starts to shine when it gets near the Sun.

constellation A group of bright stars that appear to form a pattern in the sky.

continental drift The gradual movement of the continents across the Earth.

core The centre part of a body.

corona The outer part of the Sun's atmosphere.

cosmos Another word for the universe.

crater A circular pit in the surface of a planet or moon.

crust The hard outer layer of a planet or a moon.

disk The face of a planet, which we see as a circle in telescopes.

double star A star that looks like a single star but is actually two stars close together.

eclipse When one heavenly body passes in front of another and blots out its light. An eclipse of the Sun, or a solar eclipse, takes place when the Moon passes in front of the Sun as we view it from Earth.

eclipsing binary A kind of variable star. It is a binary (two-star) system in which the stars regularly eclipse, or pass in front of, one another. This causes the brightness of the system to vary.

electromagnetic waves Waves and rays given out by the Sun.

erosion The gradual wearing away of the landscape by flowing water, the weather, wind, and so on.

evening star Usually the planet Venus appearing in the western sky just after sunset. Mercury can be an evening star, too.

expanding universe The idea that the Universe is expanding, or getting bigger.

falling star A popular name for a meteor.

fault A crack in the surface of a planet or moon caused by massive movements in the rocks.

flare A massive explosion on the Sun.

galaxy A "star island" in space. Our own galaxy is called the Milky Way.

globular cluster A globe-shaped group containing

hundreds of thousands of stars.

gravity The pull, or force of attraction, that every body has because of its mass.

greenhouse effect When the atmosphere of a planet traps heat like a greenhouse.

heavens The night sky; the heavenly bodies are the objects we see in the night sky.

ice caps Sheets of ice found at the north and south poles of Earth and Mars.

impact crater A crater made by the impact (blow) of a meteorite.

inner planets The four planets relatively close together in the inner part of the solar system — Mercury, Venus, Earth, and Mars.

interplanetary Between the planets.

interstellar Between the stars.

interstellar matter Gas and dust found between the stars.

irregulars Irregular galaxies, which have no definite shape.

latitude Of a place; how far it is away from Earth's Equator. It is measured in degrees.

lava Molten rock that pours out of volcanoes.

light-year A unit astronomers use for measuring distances in space. It is the distance light travels in a year — about 6 million million miles (10 million million kilometres).

lunar To do with the Moon.

magnetic field The region around a planet or a star in which its magnetism acts.

magnitude A measure of a star's brightness.

mantle A rocky layer beneath the crust of a rocky planet or moon.

meteor A streak of light produced when a meteoroid burns up in Earth's atmosphere.

meteorite A lump of rock from outer space that falls to the ground.

milky way A faint band of light seen in the night sky. Our galaxy is also called the Milky Way.

minor planets Another name for the asteroids.

moon The common name for a satellite.

morning star Usually the planet Venus seen in the eastern sky before sunrise. Mercury can also be a morning star.

nebula A cloud of gas and dust in space

neutron star A very dense star made up of neutrons packed tightly together.

neutrons Tiny particles found in the nuclei (centres) of most atoms.

northern hemisphere The half of the world north of the Equator. The northern celestial hemisphere is the part of the sky above the Northern Hemisphere.

nova A star that brightens suddenly and appears to be a new star. Nova means "new."

nuclear fusion A nuclear reaction in which light atoms (such as hydrogen) combine, or fuse, together. It releases enormous energy.

nuclear reaction A process that involves the nuclei (centres) of atoms.

open cluster A group of up to a few hundred stars that travel through space together.

orbit The path in space one body follows when it circles around another, such as the Moon's orbit around Earth.

outer planets The planets in the outer part of the solar system — Jupiter, Saturn, Uranus, Neptune, and Pluto.

phases The apparent changes in shape of Mercury, Venus, and the Moon in the sky. The changes come about because we see more or less of the surface of these bodies lit up by the Sun as time goes by.

photosphere The glaring surface of the Sun.

planet One of nine bodies that circle around the Sun; or more generally, a body that circles around a star.

planetary nebula A cloud of gas and dust puffed out by a dying star.

planisphere A circular device for showing the night sky on any night of the year.

plate tectonics The science that deals with the movements of pieces, or plates, of Earth's crust.

pole star Also called North Star; the star that is located in the sky almost directly above Earth's North Pole. Astronomers call it Polaris.

probe A spacecraft sent to explore other heavenly bodies, such as planets, moons, asteroids, and comets.

prominence A great fountain of hot gas that shoots out from the Sun.

pulsar A rapidly spinning tiny star, which flashes pulses of light or other radiation toward us.

quasar A body that looks like a star but is much farther away than the stars and is as bright as hundreds of galaxies.

radiation The heavenly bodies give off energy as radiation — as light rays, infrared rays, gamma rays, X-rays, ultraviolet rays, microwaves, and radio waves.

radio telescope A telescope designed to gather radio waves from the heavens.

red giant A large red star. Stars swell up to become red giants when they begin to die.

redshift A movement, or shift, in the lines in the spectrum of a star or galaxy toward the red end. This indicates that the object is moving away from us.

reflector A reflecting telescope; one that uses mirrors to gather light from the heavens.

refractor A refracting telescope; one that uses lenses to gather light from the heavens.

retrograde orbit An orbit in which a moon travels in the opposite direction of normal.

satellite A small body that orbits around a larger one; a moon. Also the usual name for an artificial satellite, an orbiting spacecraft.

shepherd moon A moon that keeps the particles of a planet's ring in place.

shooting star A popular name for a meteor.

solar To do with the Sun.

solar system The Sun and the bodies that circle around it, including planets, comets, and asteroids.

solar wind A stream of charged particles given off by the Sun.

southern semisphere The half of the world south of the Equator. The southern celestial hemisphere is the part of the sky above the Southern Hemisphere.

spectral lines Thin, dark lines in the spectrum of a star.

spectrum A colour band produced when the light from the Sun or a star is split into its separate wavelengths (colours), for example by passing it through a prism.

spirals Spiral galaxies; galaxies with a spiral shape.

star A huge ball of very hot gas, which gives off energy as light, heat, and other radiation.

sunspot A darker, cooler region of the Sun's surface.

supercluster A large grouping of clusters of galaxies.

supergiant The biggest kind of star, typically hundreds of times bigger across than the Sun.

supernova The explosion of a supergiant star.

terrestrial planets The planets that are rocky like Earth — Mercury, Venus, and Mars.

tides The regular falling and rising of ocean waters, caused mainly by the Moon's gravity.

transit The crossing of the Sun's face, as viewed from Earth, by Mercury or Venus.

universe Space and everything that is in it — galaxies, stars, planets, moons, and energy.

white dwarf A small, dense star; stars like the Sun eventually turn into a white dwarf when they die.

year The time it takes a planet to circle once around the Sun.

zodiac An imaginary band in the heavens, through which the Sun and planets appear to travel.

zone A lighter band on the face of a gassy planet like Jupiter.

Index